Principles of Design, Installation and Maintenance

Malcolm Doughton and John Hooper

CENGAGE
Learning·

Australia • Brazil • Japan • Korea • Mexico • Singapore • Spain • United Kingdom • United States

Principles of Design, Installation and Maintenance
Malcolm Doughton and John Hooper

Publishing Director: Linden Harris

Commissioning Editor: Lucy Mills

Editorial Assistant: Claire Napoli

Project Editor: Alison Cooke

Production Controller: Eyvett Davis

Marketing Executive: Lauren Mottram

Typesetter: S4Carlisle Publishing Services

Cover design: HCT Creative

Text design: Design Deluxe

For product information and technology assistance,
contact **emea.info@cengage.com.**

For permission to use material from this text or product,
and for permission queries,
email **emea.permissions@cengage.com.**

British Library Cataloguing-in-Publication Data
A catalogue record for this book is available from the British Library.

ISBN: 978-1-4080-3997-7

Cengage Learning EMEA
Cheriton House, North Way, Andover, Hampshire, SP10 5BE
United Kingdom

Cengage Learning products are represented in Canada by Nelson Education Ltd.

For your lifelong learning solutions, visit **www.cengage.co.uk**

Purchase your next print book, e-book or e-chapter at
www.cengagebrain.com

Printed in Malta by Melita Press
1 2 3 4 5 6 7 8 9 10 – 14 13 12

Dedication

This series of study books is dedicated to the memory of Ted Stocks whose original concept, and his publication of the first open learning material specifically for electrical installation courses, forms the basis for these publications. His contribution to training has been an inspiration and formed a solid base for many electricians practising their craft today.

The Electrical Installation Series

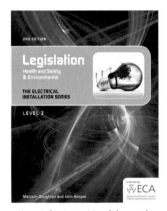

Legislation: Health and Safety & Environmental

Organizing and Managing the Work Environment

Installing Wiring Systems

Planning and Selection for Electrical Systems

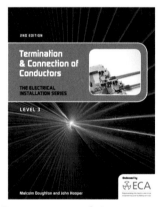

Termination and Connection of Conductors

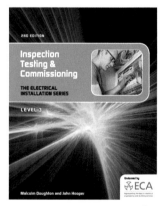

Inspection Testing and Commissioning

Fault Finding and Diagnosis

Maintaining Electrotechnical Systems

Contents

About the authors

Malcolm Doughton

Malcolm Doughton, I.Eng, MIET, LCG, has experience in all aspects of electrical contracting and has provided training to heavy current electrical engineering to HNC level. He currently provides training on all aspects of electrical installations, inspection, testing, and certification, health and safety, PAT and solar photovoltaic installations. In addition Malcolm provides numerous technical articles and is currently managing director of an electrical consultancy and training company.

John Hooper

John Hooper spent many years teaching a diverse range of electrical and electronic subjects from craft level up to foundation degree level. Subjects taught include: Electrical Technology, Engineering Maths, Instrumentation, P.L.C.s, Digital, Power and Microelectronic Systems. John has also taught various electrical engineering subjects at both Toyota and JCB. Prior to lecturing in further and higher education he had a varied career in both electrical engineering and electrical installations.

Acknowledgements

The authors and publisher would like to thank Chris Cox and Charles Duncan for their considerable contribution in bringing this series of study books to publication. We extend our grateful thanks for their unstinting patience and support throughout this process.

The authors and publisher would also like to thank the following for providing pictures for the book:
ABB Ltd
Andy Darvill
Arcol
Brook Crompton UK Ltd
Crabtree
EATON
Epcos
Finder SpA
GDC Group Ltd
Horstmann
Megger
Mira Showers
MK Electric
Newey and Eyre
Omron Electronics Ltd
OSRAM
RS Online
Shutterstock
TLC Direct
Toolstation
Vishay Intertechnology

This book is endorsed by:

Representing the best in electrical engineering and building services

Founded in 1901, the Electrical Contractors' Association (ECA) is the UK's leading trade association representing the interests of contractors who design, install, inspect, test and maintain electrical and electronic equipment and services.

www.eca.co.uk

Study guide

This studybook has been written and compiled to help you gain the maximum benefit from the material contained in it. You will find prompts for various activities all the way through the studybook. These are designed to help you ensure you have understood the subject and keep you involved with the material.

Where you see 'Sid' as you work through the studybook, he is there to help you and the activity 'Sid' is undertaking will indicate what it is you are expected to do next.

Task

Use your calculator to check this out. To do this you use the power key (x^y) on your calculator. So here we confirm 6^0, where x (the base) is 6 and y is the power 0.

Key in $\boxed{6}$ $\boxed{x^y}$ $\boxed{0}$ $\boxed{=}$ you will see the answer = 1.

Task A 'Task' is an activity that may take you away from the book to do further research either from other material or to complete a practical task. For these tasks you are given the opportunity to ask colleagues at work or your tutor at college questions about practical aspects of the subject. There are also tasks where you may be required to use manufacturers' catalogues to look up your answer. These are all important and will help your understanding of the subject.

Try this

Transpose each of the following equations for Y:

1   $YM = BD$ _____

2   $A - B = X - Y$ _____

3   $ABC = \dfrac{WX}{Y}$ _____

4   $\dfrac{A + B}{2Y} = \dfrac{E}{F}$ _____

Try this A 'Try this' is an opportunity for you to complete an exercise based on what you have just read, or to complete a mathematical problem based on one that has been shown as an example.

Remember

10^1 is $1 \times 10 = 10$

Any number raised to the power 0 is $= 1$ so $10^0 = 1$ and so is $-8^0 = 1$ and $0.125^0 = 1$ and so on.

Remember A 'Remember' box highlights key information or helpful hints.

RECAP & SELF ASSESSMENT

Circle the correct answers.

1   The decimal number 0.000 01 is equal to the fraction:

   a. $\dfrac{1}{1000}$

   b. $\dfrac{1}{10000}$

   c. $\dfrac{1}{100000}$

   d. $\dfrac{1}{1000000}$

2   The value of X in the equation $3x-7 = 5$ is:

   a. 5

   b. 4

   c. 3

   d. 6

Recap & Self Assessment At the beginning of all the chapters, except the first, you will be asked questions to recap what you learned in the previous chapter. At the end of each chapter you will find multichoice questions to test your knowledge of the chapter you have just completed.

Note

The calculations in this chapter are carried out using a Natural-VPAM calculator. Calculator functions vary; modern scientific calculators require entries in a different sequence to earlier calculators so to check your calculator enter

$\boxed{\sqrt{}}$ $\boxed{9}$ $\boxed{=}$ if the answer is 3 then your calculator is one of the current series of calculators.

Note 'Notes' provide you with useful information and points of reference for further information and material.

This studybook has been divided into Parts, each of which may be suitable as one lesson in the classroom situation. If you are using the studybook for self tuition then try to limit yourself to between 1 hour and 2 hours before you take a break. Try to end each lesson or self study session on a Task, Try this or the Self Assessment Questions.

When you resume your study go over this same piece of work before you start a new topic.

Where answers have to be calculated you will find the answers to the questions at the back of this book but before you look at them check that you have read and understood the question and written the answer you intended to. All of your working out should be shown.

At the back of the book you will also find a glossary of terms which have been used in the book.

A 'progress check' at the end of Chapter 6, and an 'end test' covering all the material in this book, are included so that you can assess your progress.

There may be occasions where topics are repeated in more than one book. This is required by the scheme as each unit must stand alone and can be undertaken in any order. It can be particularly noticeable in health and safety related topics. Where this occurs read the material through and ensure that you know and understand it and attempt any questions contained in the relevant section.

You may need to have available for reference current copies of legislation and guidance material mentioned in this book. Read the appropriate sections of these documents and remember to be on the look-out for any amendments or updates to them.

Your safety is of paramount importance. You are expected to adhere at all times to current regulations, recommendations and guidelines for health and safety.

Unit eight

Principles of design, installation and maintenance

Material contained in this unit covers the knowledge requirement for C&G Unit No. 2357-308 (ELTK 08), and the EAL unit QELTK3/008.

Unit eight considers mathematical principles and the standard units of measurement and measuring instruments. It also considers mechanical and electrical science, magnetism and electricity. Supply and distribution systems, electrical circuits, ac motors, dc machines, electrical and electronic components, lighting and heating systems are also covered in this book.

You could find it useful to look in a library or online for copies of the legislation and guidance material mentioned in this unit. Read the appropriate sections and remember to be on the lookout for any amendments or updates to them.

Before you undertake this unit read through the study guide on page vii. If you follow the guide it will enable you to gain the maximum benefit from the material contained in this unit.

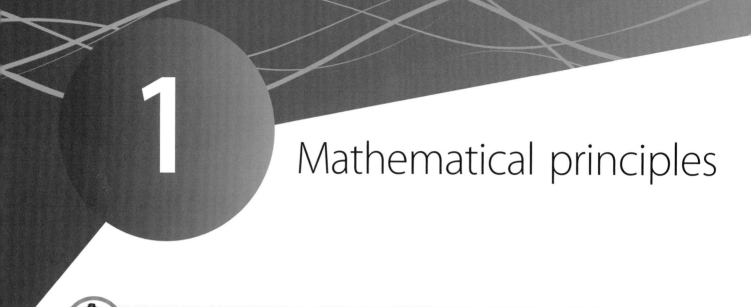

1 Mathematical principles

RECAP

Complete the following to remind yourself of some important facts that you should remember from previous studies on this subject.

A part of a whole number is called a _____.

A fraction has a number above and below a dividing line, these are the _____ and the _____.

Reducing the fraction $\frac{4}{8}$ to $\frac{1}{2}$ is usually called _____ the fraction.

The mathematical term for a whole number is an _____.

The lowest common multiple (LCM) of 4 and 5 is _____.

The reciprocal of 5 as a fraction is _____ or as a decimal is _____.

LEARNING OBJECTIVES

On completion of this chapter you should be able to:

● Identify different types of fractions and apply mathematical principles to solve fractions and percentages.

● Apply algebra to solve simple equations.

● Identify and apply laws of indices.

● Apply and manipulate powers of ten.

● Apply mathematical principles to transpose formulae.

● Apply Pythagoras' theorem and trigonometry to solve right-angled triangles.

● Use statistical methods to collect, sort, analyze and present numerical data.

Part 1 Fractions and percentages

This chapter considers some of the basic mathematical principles which may be used in electrotechnical work. This should be a revision exercise and so the subjects are not considered in great depth.

Fractions

Fractions are made up of two numbers, the numerator on the top and the denominator on the bottom.

Figure 1.1 *Fractions in all shapes and sizes*

Proper fraction

A proper fraction is one in which the numerator is less than the denominator, for example: $\dfrac{1}{2}, \dfrac{5}{8}$ and $\dfrac{3}{4}$.

Improper fraction

An improper fraction is one in which the numerator is greater than the denominator, such as: $\dfrac{9}{4}, \dfrac{16}{5}$ and $\dfrac{22}{7}$.

Mixed fraction

A mixed fraction consists of an integer (whole number) and a proper fraction such as $8\dfrac{1}{2}$.

Decimals

Decimals are fractions in which the denominator is 10, 100, 1 000, and so on.

Percentages

A percentage is a way of expressing a number as a fraction of 100 (the denominator is always 100) so $58\% = \dfrac{58}{100}$. An

Table 1.1 *Decimal fractions*

$$\frac{1}{10} = 0.1$$
$$\frac{1}{100} = 0.01$$
$$\frac{1}{1\,000} = 0.001$$
$$\frac{1}{10\,000} = 0.0001$$
$$\frac{1}{100\,000} = 0.00001$$
$$\frac{1}{1\,000\,000} = 0.000001$$
$$\frac{1}{10\,000\,000} = 0.0000001$$

example in electrical terms would be to determine a voltage drop of say 5% in a 400 V system.

Volt drop $= 400\,\text{V} \times \dfrac{5}{100} = 20\,\text{V}.$

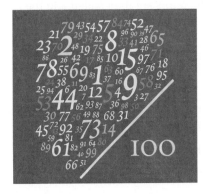

Figure 1.2 *Divided by 100 for a percentage*

Lowest common multiple (LCM) and lowest common denominator (LCD)

The lowest common multiple is the lowest number which can be divided by each of the different numbers in a group. For example the LCM of 3 and 5 is 15, because $3 \times 5 = 15$ and $5 \times 3 = 15$.

The LCM of the denominators (referred to as the lowest common denominator, LCD) of the fractions $\dfrac{1}{3}$ and $\dfrac{1}{5}$ is also 15.

So, for example, the LCD of fractions $\dfrac{1}{6} + \dfrac{4}{15}$ can be found using the following information:

Try this

Find the LCD of the fractions $\dfrac{3}{8} + \dfrac{5}{36}$

The denominators are 6 and 15 and the multiples of 6 are 6, 12, 18, 24, 30, 36 etc. The multiples of 15 are 15, 30, 45, 60 etc., and so the lowest **common** multiple value of 6 and 15 is 30.

Adding fractions

There are some simple rules when adding fractions.

1 The denominators must be converted to the lowest common denominator.
2 The numerators must be multiplied by the same multiple used for the lowest common denominator.
3 The numerators are then added and the resulting fraction simplified wherever necessary.

So for example $\dfrac{1}{3} + \dfrac{1}{5}$ is carried out as follows:

1 Determine the lowest common denominator, which in this case is 15.
2 Apply the multipliers to the numerators $\dfrac{1 \times 5}{3 \times 5} = \dfrac{5}{15}$ and $\dfrac{1 \times 3}{5 \times 3} = \dfrac{3}{15}$
3 Add the fractions $\dfrac{5}{15} + \dfrac{3}{15} = \dfrac{5 + 3}{15} = \dfrac{8}{15}$
4 Simplify the result; however in this case the fraction cannot be simplified any further.

Remember

You can't add fractions with different denominators; you need to find the lowest common denominator (LCD) for each fraction you are adding together.

Whatever you do to the bottom of a fraction, you must do the same to the top of the fraction.

Figure 1.3 *Fractions occur in everyday work*

Subtracting fractions

The same principle applies to subtracting fractions except we are subtracting rather than adding. So, for example:

$$\frac{1}{2} - \frac{1}{6} - \frac{1}{12} = \frac{6}{12} - \frac{2}{12} - \frac{1}{12} = \frac{6 - 2 - 1}{12} = \frac{3}{12} = \frac{1}{4}$$

Try this

1 $\dfrac{1}{3} + \dfrac{2}{5}$

2 $\dfrac{1}{5} + \dfrac{1}{6} + \dfrac{1}{15}$

3 $\dfrac{17}{18} - \dfrac{4}{9} - \dfrac{1}{8}$

Part 2

Multiplying fractions

There are some simple rules when multiplying fractions:

1 Multiply the numerators.
2 Multiply the denominators.
3 Simplify the fraction (if possible).

So for example $\dfrac{1}{3} \times \dfrac{9}{16} = \dfrac{1 \times 9}{3 \times 16} = \dfrac{9}{48} = \dfrac{3}{16}$

Dividing fractions

Similarly, the rules to follow when dividing fractions are as follows:

1 Turn the second fraction upside down (this is now the reciprocal).
2 Multiply the first fraction by that reciprocal.
3 Simplify the fraction (if possible).

So for example $\dfrac{1}{2} \div \dfrac{1}{6} = \dfrac{1}{2} \times \dfrac{6}{1} = \dfrac{1 \times 6}{2 \times 1} = \dfrac{6}{2} = 3$

 Try this

1 $\dfrac{3}{4} \times \dfrac{7}{11}$ _____

2 $\dfrac{1}{2} \times \dfrac{3}{4} \times \dfrac{5}{7}$ _____

3 $\dfrac{4}{5} \div \dfrac{3}{15}$ _____

4 $\dfrac{4}{15} \div \dfrac{9}{10}$ _____

Adding and subtracting mixed fractions

As you may have guessed there are some simple rules when adding and subtracting mixed fractions.

Adding:

1 Convert to improper fractions.
2 Then add them (using addition of fraction methods previously covered).
3 Convert the result back to a mixed fraction.

So, for example:

$$6\dfrac{3}{4} + 3\dfrac{5}{8} = \dfrac{27}{4} + \dfrac{29}{8} = \dfrac{54}{8} + \dfrac{29}{8}$$

$$= \dfrac{54 + 29}{8} = \dfrac{83}{8} = 10\dfrac{3}{8}$$

Subtracting mixed fractions

Follow these simple steps:

1 Convert to improper fractions.
2 Then subtract them (using subtraction of fraction methods previously covered).
3 Convert the result back to a mixed fraction.

So, for example:

$$15\dfrac{3}{4} - 8\dfrac{5}{6} = \dfrac{63}{4} - \dfrac{53}{6} = \dfrac{189}{12} - \dfrac{106}{12}$$

$$= \dfrac{83}{12} = 6\dfrac{11}{12}$$

Try this

1. $6\dfrac{4}{5} + 9\dfrac{1}{2}$ _____

2. $6\dfrac{1}{8} - 2\dfrac{3}{4}$ _____

3. $5\dfrac{1}{3} - 2\dfrac{2}{3}$ _____

Multiplying mixed fractions

Using the information we have gained so far we need to follow these steps:

1   Convert to improper fractions.

2   Multiply fractions.

3   Convert the result back to a mixed fraction.

Example:

$$1\frac{5}{9} \times 2\frac{1}{7} = \frac{14}{9} \times \frac{15}{7} = \frac{210}{63}$$

We can simplify this example by dividing top and bottom by 21 so

$$\frac{210}{63} = \frac{10}{3} = 3\frac{1}{3}$$

Decimals and fractions

Convert fractions to decimal numbers.

To convert π **(pi)** or $\dfrac{22}{7}$ to a decimal number we simply divide the numerator by the denominator. We can do this using a calculator giving us 3.1428 and to three decimal places this would be 3.143.

Figure 1.4 *Pi*

Try this

1   $2\dfrac{2}{3} \times 4\dfrac{1}{2}$ _____

2   $1\dfrac{5}{7} \times 1\dfrac{5}{9}$ _____

Try this

Convert the following fractions to decimal numbers correct to three decimal places.

1   $\dfrac{43}{7}$ _____

2   $\dfrac{67}{6}$ _____

Fractions and percentages

To convert a fraction into a percentage the two simple steps are: 1. divide the numerator by the denominator, (to convert the proper fraction to a decimal fraction). 2. multiply by 100 to represent the decimal fraction as a percentage.

So, to convert $\frac{3}{8}$ into a percentage:

1 $\frac{3}{8} = 0.375$

2 $0.375 \times 100 = 37.5\%$.

To convert a percentage into a fraction the steps for whole number percentages are simply to divide the percentage by 100 and then simplify the result wherever possible.

So 75% as a fraction is $\frac{75}{100} = \frac{3}{4}$

Where the percentage is not a whole number then an addition step is required. For example, to convert 62.5% to a fraction $\frac{62.5}{100}$ the numerator is not a whole number and to correct this we need to multiply both the numerator and the denominator by 10 so it becomes:

$$\frac{625 \div 25}{1000 \div 25} = \frac{25 \div 5}{40 \div 5} = \frac{5}{8}$$

(we divide the numerator and denominator by 25 and again by 5 to simplify the fraction).

Using a scientific calculator to work out percentages

To determine 3% of 230V:

Key in $\boxed{2}\boxed{3}\boxed{0}\ \boxed{\times}\ \boxed{3}\ \boxed{\%}$ and 6.9 is displayed.

Answer = 6.9 V.

Note

The calculations in this chapter are carried out using a Natural-VPAM calculator. Calculator functions vary; modern scientific calculators require entries in a different sequence to earlier calculators so to check your calculator enter $\boxed{\sqrt{}}\ \boxed{9}\ \boxed{=}$ If the answer is 3 then your calculator is one of the current series of calculators.

Remember

Mathematical operations need to be carried out in the correct sequence and the term BODMAS may be useful in helping to remember the correct order. It stands for Brackets, power Of, Division, Multiplication, Addition and Subtraction.

So is the correct answer for $2 + 3 \times 7$ either 35 or 23? Using the above reference the correct answer will be $3 \times 7 = 21$ then $21 + 2 = 23$.

Try this

Convert the following to fractions:

1 87.5% _____

2 12.5% _____

Convert the following to percentages:

1 $\frac{5}{8}$ _____

2 $\frac{11}{16}$ _____

Part 3

Algebra

Algebra is a branch of mathematics which employs the use of symbols or letters, either upper or lower case, to solve mathematical problems.

$$P = VICos\Phi \quad PF = Cos\Phi = \frac{kW}{kVA} \quad PF = Cos\Phi = \frac{R}{Z}$$

$$P = I^2R \quad P = \frac{V^2}{R} \quad I = \frac{V}{Z} \quad \frac{V_P}{V_S} = \frac{N_P}{N_S} = \frac{I_S}{I_P} \quad P = VI$$

$$E = BIv \quad F = BII \quad R = \frac{\rho l}{A} \quad R_T = R_1 + R_2 + R_3$$

$$Q = It \quad X_L = 2\Pi fL \quad X_C = \frac{1}{2\Pi fC} \quad Q = CV$$

Figure 1.5 *Algebra is used to solve many electrical problems*

Remember

In algebra the multiplication sign is usually left out as this avoids confusion when the letter x is used for a variable. So, for example a × b is shown as 'ab' and 3 × y is shown as '3y'.

Addition and subtraction

The simple rules of transposition to note are:

- Move the symbol required to the left hand side (LHS) of the equation and move all others to the right hand side (RHS).
- When a symbol moves from one side of the = sign to the other side, it changes sign.

To determine the value of X in the equation $V + W = X - Y$ then:

$$-X + V + W = -Y \text{ and so } -X = -Y - V - W$$

However, we need X, so simply change its sign and do the same to the RHS of the equation, so $X = Y + V + W$.

Multiplication and division

The rules of transposition to note here are:

- Move the symbol required across the = sign so it is on the top LHS of the equation.
- Move all other symbols away from it, across the = sign but in the opposite position, from top to bottom or vice versa.
- Signs are not changed.

To find the value of S when $\dfrac{WXY}{Z} = \dfrac{RS}{T}$

Leave S where it is on the top line and move R and T and so the formula becomes:

$$\frac{WXYT}{ZR} = S, \text{ which is the same as } S = \frac{WXYT}{ZR}$$

More complex formulae such as $\dfrac{a(b + c)}{xy} = \dfrac{d}{t}$ transposed to find t require a little more thought. Moving 't' to top LHS gives $\dfrac{ta(b + c)}{xy} = d$ and to get t on its own we need to move the rest of the LHS which gives $t = \dfrac{dxy}{a(b + c)}$.

Where $\dfrac{a(b + c)}{xy} = \dfrac{d}{t}$ to determine what 'c' equals we first leave the expression (b + c) on the top line, since 'c' is part of that expression.

So $(b + c) = \dfrac{dxy}{at}$; to remove the LHS bracket we must bracket the RHS so $b + c = \left(\dfrac{dxy}{at}\right)$ so $c = \left(\dfrac{dxy}{at}\right) - b$.

Simple equations

To solve the equation $5a - 6 = 3a - 8$ to determine the value of 'a' we first subtract 3a from both sides of the equation so $5a - 6 - 3a = -8$ and so $2a - 6 = -8$ which then gives $2a = -8 + 6$ so $2a = -2$ and therefore $a = -1$.

To check the answer we put the value into the formula so $5(-1) - 6 = 3(-1) - 8$.

Simplified to $-5 - 6 = -3 - 8$ and so $-11 = -11$.

Try this

Transpose each of the following equations for Y:

1 $YM = BD$ _____

2 $A - B = X - Y$ _____

3 $ABC = \dfrac{WX}{Y}$ _____

4 $\dfrac{A + B}{2Y} = \dfrac{E}{F}$ _____

Try this

Find the value of x in each of the following equations and check your answers by substitution:

1 $X + 6 = 15$ _____

2 $6X + 4 = 40$ _____

3 $4X - 6 = 12X - 30$ _____

Part 4

Indices (powers of 10)

Indices are a shorthand way of writing very large and very small numbers.

The 'power of ten' means how many times 1 is multiplied by 10, so 10^3 is the shorthand way of writing:

$$1 \times 10 \times 10 \times 10 \text{ or } 1\,000.$$

Remember

10^1 is $1 \times 10 = 10$

Any number raised to the power 0 is = 1. Therefore $10^0 = 1$ and $-8^0 = 1$ and $0.125^0 = 1$ and so on.

Task

Use your calculator to check this out. To do this you use the power key (x^y) on your calculator. So here we confirm 6^0, where x (the base) is 6 and y is the power 0.

Key in 6 x^y 0 = you will see the answer = 1.

The exponent (EXP or x•) button on your calculator allows you to enter numbers to powers of 10. When using this button on your calculator always key in the number then press the EXP or x• button and key in the power (exponent) number. Also remember the (–) button if the power (exponent) is negative.

In Table 1.2 the value of 10^3 should be read as 'ten to the power of three' and 10^{-3} as 'ten to the power of minus three', which is another way of putting $\frac{1}{10^3}$.

Table 1.2 *Indices (those marked * are commonly used in electrical installation work)*

To the power	Actual Number	Title
10^{12}	1 000 000 000 000	T (tera)
10^{9}	1 000 000 000	G (giga)
10^{6}	1 000 000*	M (mega)
10^{3}	1 000*	k (kilo)
10^{2}	100	h (hecto)
10^{1}	10	d (deka)
10^{0}	1	
10^{-1}	0.1	d (deci)
10^{-2}	0.01	
10^{-3}	0.001*	m (milli)
10^{-6}	0.000 001*	μ (micro)
10^{-9}	0.000 000 001	n (nano)
10^{-12}	0.000 000 000 001	p (pico)

Transposition of formulae

It is often necessary to transpose (or rearrange) a formula in order to solve a problem. Up to now we have been carrying out simple transposition of algebraic equations. We are now going to methodically transpose formulae where the letters represent symbols for electrical and mechanical variables.

Let us consider the formula: $V = I \times R$

Where:

V is the voltage in volts (V)
I is the current in amperes (A)
R is the resistance in ohms (Ω).

V is the subject of this formula when it is the 'variable' we want to find. If we wish to determine current then the formula will need to be transposed much as we did earlier in this chapter.

$$V = IR \therefore \frac{V}{R} = I$$

Often when installing a motor we know certain information but may need to determine one particular feature for a particular application. The formula which relates power to torque is $P = \frac{2\pi NT}{60}$

Where:

P is the power in watts (W)
N represents the rotational speed in revs/min (rpm)
T represents the torque in Newton-metres (Nm).

To determine the torque produced we need to make T the subject of the formula.

$$P = \frac{2\pi NT}{60} \text{ so } T = \frac{60P}{2\pi N}.$$

The impedance (Z) of an ac circuit containing resistance (R) and (L) is given by the formula $Z = \sqrt{R^2 + X_L^2}$.

To make the inductive reactance X_L the subject of the formula, first we must square both sides $\mathbf{Z^2} = \sqrt{(\mathbf{R^2} + \mathbf{X_L^2})^2}$ which results in $Z^2 = R^2 + X_L^2$ as the square and the square root of the bracket have cancelled each other out. We can then make X_L^2 the subject as $X_L^2 = Z^2 - R^2$ then we need to square root both sides $\sqrt{Z^2 - R^2} = \sqrt{X_L^2}$ and the square root cancels the square on the RHS giving $X_L = \sqrt{Z^2 - R^2}$.

When we transpose formulae which include brackets the steps are similar but we need to remember the correct process and sequence.

For example, if we are to transpose the formula $P = \frac{2\Pi rN(F_1 - F_2)}{60}$ to make F_1 the subject then:

$$60P = 2\Pi rN(F_1 - F_2) = \frac{60P}{2\Pi rN} = (F_1 - F_2)$$

To remove the brackets we must bracket the LHS to give $\left(\frac{60P}{2\Pi rN}\right) = F_1 - F_2$ and so $F_1 = \left(\frac{60P}{2\Pi rN}\right) + F_2$

Try this

Make F_2 the subject of the formula $P = \dfrac{2\Pi r N (F_1 - F_2)}{60}$

Part 5 Statistics

Statistics is the study of collecting, sorting, analyzing and presenting data which is obtained largely by two methods:

1   by counting – for example, the number of consumer units sold by an electrical wholesaler over equal periods of time
2   by measurement – for example, the heights of a group of electricians.

Both methods may be used. For example, in a car factory they will be counting the number (quantity) of cars being manufactured in a given time period and also measuring the quality of the cars and their component parts before they leave the factory. All this statistical information will then be recorded and stored on their computer databases.

In electrical installation work, statistics are often presented in tables or graphs for ease of reference.

Tables, charts and diagrams

An electrical contractor will employ a number of people in various jobs within the company. The table below gives typical employee numbers for a fairly large electrical contracting company.

Table 1.3 *Typical employee statistics*

Type of personnel	Number employed	Percentage
Electricians	130	65
Clerical staff	30	15
Apprentices	20	10
Draughtsmen	10	5
Labourers	10	5
Total	200	100

This data can be represented pictorially in several ways and these include:

Pie charts, in which the area of a circle represents the whole and the areas of the sectors (slices) of the circle are made proportional to the parts which make up the whole.

Horizontal bar charts, having data represented by equally spaced horizontal rectangles (bars), and

Vertical bar charts, in which data are represented by equally spaced vertical rectangles (bars).

Pie chart

To produce a pie chart to show the data in Table 1.3 we first need to draw a circle of any suitable radius; you will need a protractor, then the angles of the sectors are drawn using a protractor.

The whole, 200 employees, corresponds to 360° and the angle of each sector (proportional to each part) can be calculated like this:

130 employees corresponds to $360 \times \dfrac{130}{200} = 234°$ (Electricians)

30 employees corresponds to $360 \times \dfrac{30}{200} = 54°$ (Clerical Staff)

and so on, giving the angles at the centre of the circle for each group of employee numbers as: 18°, 18°, 36°, 54° and 234° respectively.

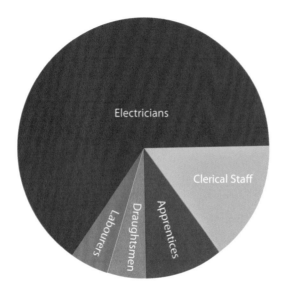

Figure 1.6 *Pie chart for the data in Table 1.3*

Remember

Tables, charts and diagrams are widely used to give a clear picture of statistical information.

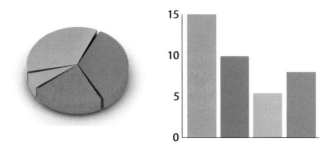

Figure 1.7 *Pie and bar charts*

Horizontal and vertical bar charts

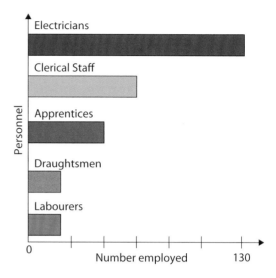

Figure 1.8 *Horizontal bar chart for the data in Table 1.3*

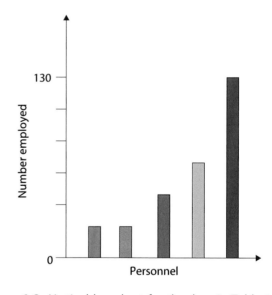

Figure 1.9 *Vertical bar chart for the data in Table 1.3*

Part 6

Use of brackets

Remember

Brackets indicate that all the terms, or quantities, inside the brackets must be operated upon by the term, or quantity, outside the brackets when the brackets are removed.

To remove the brackets simply multiply the terms inside the brackets by the terms outside the brackets.

Multiplying or expanding brackets

1   If the brackets are multiplied by a positive (+) quantity the sign of the terms within the brackets remain unchanged when the brackets are removed.

2   If the brackets are multiplied by a negative (–) quantity all the signs within the brackets must be changed, from + to – or – to +, on removal of the brackets.

When we need to expand and simplify an expression such as:

$$3(a + b) - 2(3a - 4b) = 0$$

this will become: $3a + 3b - 6a + 8b = 0$

Then simplifying this equation it becomes: $-3a + 11b = 0$.

Putting this into a more standard format we get: $11b - 3a = 0$.

This process has changed $3(a + b) - 2(3a - 4b) = 0$ to become $11b - 3a = 0$.

Try this

Expand and simplify, where possible, the following expressions:

1   $l(R + r)$ _____

2   $7(y - 3) - 2(3 + y)$ _____

3   $4(p + 2q) - 2(p - 3q)$ _____

Indices

Squares

When a number is multiplied by itself the result is the **square** of the number.

For example $3 \times 3 = 3^2$

Using your calculator enter $\boxed{3}$ $\boxed{\times}$ $\boxed{3}$ $\boxed{=}$ 9.

3^2 is called 'three to the power two' or 'three squared'.

Cubes

If a number is multiplied by itself three times the result is the cube of the number.

For example $5 \times 5 \times 5 = 5^3$

5^3 is called 'five to the power of three' or 'five cubed'.

Using your calculator enter

5 × 5 × 5 = 125

Square roots

Figure 1.10 *Square root sign*

Remember

Calculator functions vary; modern scientific calculators require entries in a different sequence to earlier calculators so to check your calculator:

Enter √ 9 = . If the answer is 3 then your calculator is one of the current series of calculators.

Earlier calculators required a sequence of 9 √ = to give an answer of 3.

Whilst 49 is the square of 7 it is also true that 7 is the square root of 49. So $7^2 = 49$ and $7 = \sqrt{49}$.

To find $\sqrt{1024}$ on your calculator you will need to enter.

√ 1 0 2 4   answer is 32

Task

Check your calculator function before you continue.

Laws of indices

Law 1: When multiplying quantities together containing powers having the same base, add the powers.

So for example $3^2 \times 3^5 = 3^{2+5} = 3^7$

Let's evaluate 3^7 on your calculator.

Enter 3 X^y 7 = 2187

Law 2: When dividing quantities containing powers having the same base, subtract the powers.

So for example $\dfrac{5^4}{5^2} = 5^{4-2} = 5^2$

Law 3: When a quantity containing a power is raised to a further power, multiply the powers together.

So for example $(6^4)^3 = 6^{4\times3} = 6^{12}$

Negative indices

It is often necessary to express an index as a negative quantity.

Let's compare the following two statements:

$$\frac{1}{3^4} \times 3^6 = \frac{3^6}{3^4} = 3^{6-4} = 3^2 \qquad \text{(Law 2)}$$

$$3^{-4} \times 3^6 = 3^{6-4} = 3^2 3^{-4} \times 3^6 \qquad \text{(Law 1)}$$

We can see from this that $\dfrac{1}{3^4} = 3^{-4}$

Try this

1 **Find the square root of the following:**
a) 289 _____

b) 155236 _____

2 **Find the cube of:**
a) 9 _____

b) 31 _____

3 **Simplify:**
a) $7^2 \times 7^4 \times 7$ _____

b) $(9^2)^3$ _____

4 **Express the following with negative indices:**
a) 10^4 _____

b) $\dfrac{1}{10^7}$ _____

Part 7

Standard form

Standard form is used for very large or very small numbers. A number is in standard form when there is only one digit to the left of the decimal point. So, for example 45 750 000 in standard form is achieved by moving the decimal point 7 places to the left. This is by increasing the power of 10 to 7 and so it becomes 4.575×10^7; similarly $268\ 500 = 2.685 \times 10^5$.

A number such as 0.000 000 004 8 in standard form requires the decimal point to be moved 9 points to the right and so it becomes 4.8×10^{-9}.

 Try this

Change the following numbers into standard form:

1 978 000 000 _____

2 0.000 003 45 _____

3 $\dfrac{1}{1\ 000}$ _____

Pythagoras' theorem

Pythagoras' theorem states that in any right-angled triangle the square on the hypotenuse is equal to the sum of the squares on the other two sides. The hypotenuse is the side opposite the right angle.

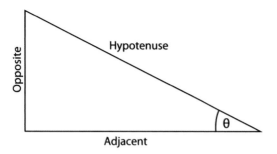

Figure 1.11 *Sides of a right-angled triangle*

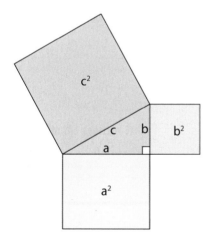

Figure 1.12 *Pythagoras' theorem.*

We can see from Figure 1.12 that the area of the square on side **c** = the area of the square on side **a** plus the area of the square on side **b**.

In algebraic terms, $c^2 = a^2 + b^2$, and from our previous examples we know that the length of the hypotenuse may be found by $c = \sqrt{a^2 + b^2}$ providing the lengths of sides a and b are known.

Also, $a = \sqrt{c^2 - b^2}$ and $b = \sqrt{c^2 - a^2}$.

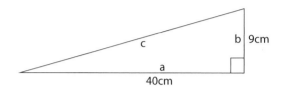

Figure 1.13 *Right-angled triangle*

To determine the length of the hypotenuse of the triangle in Figure 1.13 we can use the formula:

$$c = \sqrt{a^2 + b^2}$$
$$= \sqrt{40^2 + 9^2}$$
$$= \sqrt{(1600 + 81)}$$
$$= \sqrt{1681}$$
$$= 41\text{cm}$$

To use your calculator enter

$\boxed{\surd}$ $\boxed{4}$ $\boxed{0}$ $\boxed{X^2}$ $\boxed{+}$ $\boxed{9}$ $\boxed{X^2}$ $\boxed{=}$ 41

As we can see the answer is 41cm

Try this

Find the missing lengths in the following triangles.

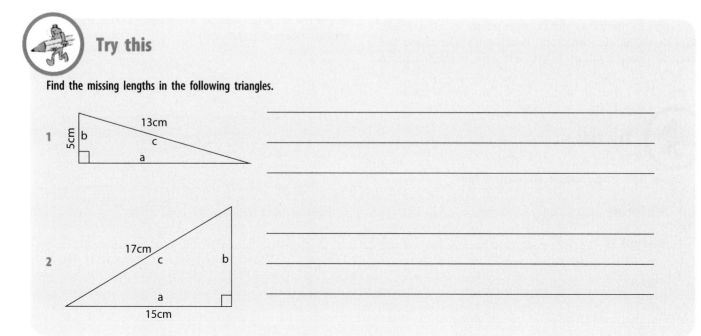

Part 8

Trigonometry

Trigonometry is the study of the relations between the **sides** and the **angles** of **triangles**.

The **hypotenuse** is the side opposite to the right angle.

The opposite side is so called because it is opposite to the angle θ (theta).

The adjacent side is the side adjacent to angle θ.

Trigonometric ratios

These ratios are expressed in terms of sine, cosine and tangent and they are abbreviated sin, cos and tan respectively.

For the angle θ the ratios are:

$$\sin\theta = \frac{\text{opposite}}{\text{hypotenuse}}$$

$$\cos\theta = \frac{\text{adjacent}}{\text{hypotenuse}}$$

$$\tan\theta = \frac{\text{opposite}}{\text{adjacent}}$$

The values of sin, cos and tan can be obtained from any scientific calculator.

Since we are dealing with a right-angled triangle these values (expressed in degrees) will range between 0° and 90°.

Now let's draw the simplest right-angled triangle (a 3:4:5 triangle) and find the sine, cosine and tangent of the angle by using the trigonometric ratios.

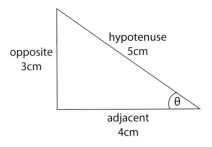

$$\sin\theta = \frac{\text{opposite}}{\text{hypotenuse}} = \frac{4}{5} = 0.6$$

$$\cos\theta = \frac{\text{adjacent}}{\text{hypotenuse}} = \frac{4}{5} = 0.8$$

$$\tan\theta = \frac{\text{opposite}}{\text{adjacent}} = \frac{3}{4} = 0.75$$

Now let's find the angle from your calculator. On most calculators the $\sin^{-1}$ function (the angle whose sine is) requires the shift button followed by $\sin^{-1}$ to correctly function. So we need to enter:

| shift | $\sin^{-1}$ | 0 | · | 6 | = | 36.86989765

Therefore $\sin^{-1} 0.6 = 36.87°$.

Enter:

| shift | $\cos^{-1}$ | 0 | · | 8 | = | 36.86989765.

Therefore $\cos^{-1} 0.8 = 36.87°$.

Enter:

| shift | $\tan^{-1}$ | 0 | · | 7 | 5 | = | 36.86989765

Therefore $\tan^{-1} 0.75 = 36.87°$.

So the angle θ is 36.87°.

Conversely if you know the angle θ you can obtain the sine, cosine and tangent of the angle θ from your calculator.

To use your calculator enter:

| sin | 3 | 6 | · | 8 | 6 | = | 0.600001

Therefore sin 36.87° = 0.6

| cos | 3 | 6 | · | 8 | 6 | = | 0.79999

Therefore cos 36.87° = 0.8

| tan | 3 | 6 | · | 8 | 6 | = | 0.75000

Therefore tan 36.87° = 0.75

So what use is this in our electrical installation work? Let's consider a situation where we have to run an overhead cable across a square tank which is filled with water. We can determine the cable length required using the length of the tank sides or angles.

For example

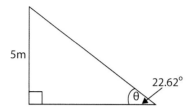

Determine the:

a)   sine of the angle θ.

b)   length of the hypotenuse.

a)   Using your calculator enter

$$\boxed{sin}\ \boxed{2}\ \boxed{2}\ \boxed{\cdot}\ \boxed{6}\ \boxed{2}\ \boxed{=}\quad 0.38461$$

Therefore sin 22.62° = 0.3847.

b) $\text{Sin } \theta = \dfrac{\text{opp}}{\text{hyp}}$ and so $\text{hyp} = \dfrac{\text{opp}}{\text{sin}} = \dfrac{5}{0.3847} = 13\text{m}$ therefore the hypotenuse = 13m.

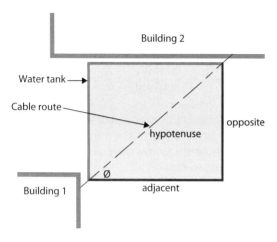

Figure 1.14 *Determining cable length*

Try this

1   Find:

a)   the cosine of the angle θ _____

b)   the length of the adjacent side _____

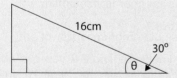

2   Find:

a)   the tangent of the angle θ _____

b)   the length of the opposite side _____

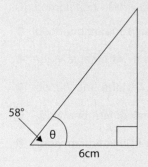

Congratulations you have now completed this first chapter. Correctly complete the self-assessment before you progress to the next chapter.

SELF ASSESSMENT

Circle the correct answers.

1 The decimal number 0.000 01 is equal to the fraction:

a. $\frac{1}{1\,000}$

b. $\frac{1}{10\,000}$

c. $\frac{1}{100\,000}$

d. $\frac{1}{1\,000\,000}$

2 The value of x in the equation $3x-7=5$ is:

a. 5

b. 4

c. 3

d. 6

3 $5^{-4} \times 5^{6}$ equals:

a. 5^{-12}

b. 5^{2}

c. 5^{-10}

d. 5^{-2}

4 To find R the formula $P=I^{2}R$ is transposed to:

a. $R=\dfrac{P}{I^{2}}$

b. $R=\dfrac{I^{2}}{P}$

c. $R=P\times I^{2}$

d. $R=\sqrt{PI}$

5 Using Pythagoras' theorem, the value of V_T is:

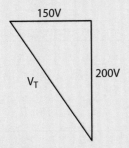

a. 400V

b. 50V

c. 350V

d. 250V

2

Standard units of measurement and measuring instruments

RECAP

Before you start work on this chapter, complete the exercise below to ensure that you remember what you learned earlier.

1 Find the LCD of the fractions $\dfrac{3}{80}$ and $\dfrac{5}{360}$

_____

_____

2 Transpose a – b = d – e to make e the subject

_____

_____

3 Convert 10^{-5} to a proper fraction

_____

_____

4 For the figure below find the:

a) tangent of the angle θ _____

b) length of the opposite side to 1 decimal place _____ .

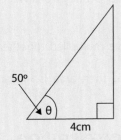

_____

_____

_____

_____

LEARNING OBJECTIVES

On completion of this chapter you should be able to:

- Identify the base units and derived units of the SI system.

- Name the symbols and factors appropriate to multiples and submultiples of SI units.

- Use SI units for the measurement of general variables.

- Identify and determine values of basic SI units which apply to electrical variables.

- Identify appropriate electrical instruments for the measurement and calculation of electrical values.

Part 1 SI units

The SI system (International System of Units) is the metric system of measurement. SI units are gradually replacing imperial units, such as yards, feet and inches for measuring length. Although we also now use the metric system there are instances where we still use imperial measures.

The use of SI units for the measurement of general variables

The SI unit for length is the metre. The unit's symbol is m and the variable symbol (as used in equations) is l. There are seven base and two supplementary units in the International System of Units. The base units you need to know about at this stage are shown in Table 2.1.

Table 2.1 *Base units*

Variable	Unit	Unit symbol	Variable symbol
Length	metre	m	l
Mass	kilogram	kg	m
Time	second	s	t
Electric current	ampere	A	I
Temperature*	kelvin	K	t

Temperature

Temperature is simply the degree of hotness or coldness of a body or environment. Although the SI unit is the kelvin it has been internationally agreed that the degree Celsius (°C) will be the unit for everyday temperature measurement. The kelvin starts at absolute zero; it is the fraction 1/273.15 of the thermodynamic temperature of the triple point of water. The degree Celsius is the kelvin minus 273.15, so 0 °C corresponds to 273.15 K. The intervals are the same for both.

Conversion of temperature units

Celsius (°C) = 5/9 (°F – 32) and Fahrenheit (°F) = 9/5 × °C +32

Example: Convert 68°F to Celsius. Celsius = 5/9 (68 – 32) = 5/9 × 36 = 20°C

Derived units

Other units are derived from the base units.

The unit of force, the newton (unit symbol N, variable symbol F), is mass multiplied by acceleration. (Acceleration is speed divided by time and speed is distance divided by time.) 1 N = 1 kg m/s^2

Area is the product of two lengths multiplied together.

As length is measured in metres then area is in square metres: m^2.

Speed, velocity and acceleration

Speed is how fast an object is going or how far the object can go in a certain amount of time. Speed takes no account of direction and is therefore a scalar quantity. A scalar quantity has magnitude only.

Example: The top speed of a Tornado F3 Jet is 2695 kilometres per hour (km/h) or 1674 mph.

Velocity is a measure of the speed of an object in a given direction; therefore velocity is a vector quantity. A vector quantity has both magnitude and direction.

Example: The velocity of the Tornado F3 Jet is 2200 km/h in a south east direction.

The SI unit for both speed and velocity is metres per second (m/s), since distance is measured in metres (m) and time is measured in seconds (s). In the UK we still use the imperial measure for speed (mph) but worldwide the km/h is the most commonly used unit for speed.

Acceleration is the change in speed or velocity over a period of time. The SI unit of acceleration is metre per second per second, m/s/s (m/s^2).

Example: If a racing car's starting velocity is 0 metres per second (0 m/s) and its final velocity is 30 m/s, the difference is 30 m/s. If this change took 3 seconds then:

$$\text{Acceleration} = \frac{30}{3} = 10 \text{ m/s}^2$$

Try this

Convert 10°C to Fahrenheit. _____

Some common variables and their SI units are shown in Table 2.2.

Table 2.2 *Some common variables, units and symbols*

Variable	Unit	Unit symbol	Variable symbol
acceleration	metre/second^2	m/s^2	a
area (cross sectional area)	square metre	m^2	A
capacitance	farad	F	c
charge (quantity of electricity)	coulomb	C	Q
density	kilogram/cubic metre	kg/m^3	ρ
electromotive force (e.m.f.)	volt	V	V
electrical energy	joule	J	W
	kilowatt hour	kWh	
force	newton	N	F
frequency	hertz	Hz	f
inductance	henry	H	L
potential difference	volt	V	V
power	watt	W	P
resistance	ohm	Ω	R
resistivity	ohm metre	Ωm	ρ
speed (velocity)	metre/sec	m/s	v
volume	cubic metre	m^3	V
work	newton metre	Nm	W
magnetic flux	weber	Wb	Φ
magnetic flux density	tesla	T	B

Multiples and submultiples of units

Sometimes it is more convenient to use multiples and submultiples of units (Table 2.3).

Table 2.3 *Common multiples and submultiples*

Prefix	Symbol	Value	
tera	T	10^{12} or	1 000 000 000 000
giga	G	10^9	1 000 000 000
mega	M	10^6	1 000 000
kilo	k	10^3	1 000
deci	d	10^{-1}	0.1
centi	c	10^{-2}	0.01
milli	m	10^{-3}	0.001
micro	μ	10^{-6}	0.000001
nano	n	10^{-9}	0.000000001
pico	p	10^{-12}	0.000000000001

The most commonly used powers of ten in electrical units are as follows:

megawatt: 1 megawatt = 1 000 000 watts, and so 1MW = 10^6 W

kilowatt: 1 kilowatt = 1000 watts, and so 1kW = 10^3 W

milliampere: 1 ampere = 1000 milliamperes and so 1A = 1000mA or 1mA = 10^{-3} A

microfarad: 1 farad = 1 000 000 microfarads and so 1 F = 1 000 000 μF or 1 μF = 10^{-6} F.

Remember

A variable has both a number and a unit.

Try this

If the velocity of a car changes from 5m/s to 15m/s in 2 seconds, what is the car's average acceleration?

_____

_____

_____

Try this

Which symbols do these units represent?

mm	*Millimetre*
mA	_____
GW	_____
kV	_____
pF	_____

Part 2

Area

The measurement of area is expressed in m^2 (square metres) and the common submultiples are mm^2, m^2 and cm^2. In a rectangular shape the area is found by multiplying the length by the breadth.

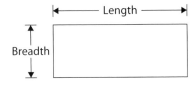

Figure 2.1 *Area of rectangle*

Example: Calculate the total cross-sectional area (csa) of trunking whose dimensions are 70mm × 50mm.

$$area = length \times breadth$$
$$= 70 \times 50mm^2$$
$$= 3500mm^2$$

The area of other shapes is calculated by using appropriate formulae.

The area of a circle is calculated by the formula:

$$\text{Area} = \pi r^2 \text{ or } \frac{\pi d^2}{4}$$

where *r* is the radius of the circle, *d* is the diameter of the circle (and when a number is squared it is multiplied by itself) and π (pi) has a value of approximately $^{22}/_7$ or 3.142.

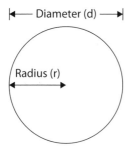

Figure 2.2 *Area of a circle*

Example: What is the area of a circle of diameter 6m?

$$\text{Area} = \pi r^2 = \frac{22 \times 3^2}{7} = 28.31\text{m}^2 \text{ or } 3.142 \times 9 = 28.28\text{m}^2$$

The area of a triangle is calculated by the formula

$$\text{area} = \frac{1}{2} \text{ base} \times \text{height}$$

Note that the height is the perpendicular height.

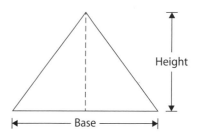

Figure 2.3 *Area of a triangle*

Example: Calculate the area in cm^2 of a triangle with a base of 2cm and a height of 4cm.

$$\text{Area} = \frac{1}{2}(2 \times 4) = 4\text{cm}^2$$

Volume

Volume is measured in m^3 (cubic metres). You may also find measurements in the submultiples of cm^3 and mm^3.

The volume of a cuboid, and in fact any regular (parallel-sided) solid, is the area of one end multiplied by the length. It is found by using the formula

$$\text{volume} = \text{length} \times \text{breadth} \times \text{depth (or height)}$$

Try this

1 Calculate the cross-sectional area, in mm^2, of 75mm × 25mm trunking.

_____

_____

2 Calculate the overall cross-sectional area, in mm^2 of a cable of nominal overall diameter of 3.5mm and π = 3.142

_____

_____

_____

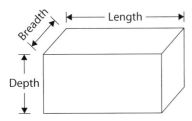

Figure 2.4 *Volume of a cuboid*

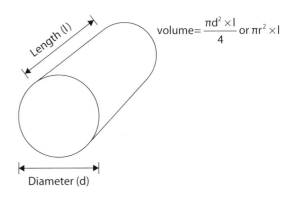

$$volume = \frac{\pi d^2 \times l}{4} \text{ or } \pi r^2 \times l$$

Figure 2.5 *Volume of a cylinder*

Example: Calculate the volume of a room 6 metres × 5 metres × 2 metres.

$$\begin{aligned} \text{volume} &= \text{length} \times \text{breadth} \times \text{depth} \\ &= 6 \times 5 \times 2 \\ &= 60\text{m}^3 \end{aligned}$$

The volume of a cylindrical object is the area of the circle multiplied by the height (or length).

Example: Find the volume of a cylindrical tank 2 metres high and of diameter 0.5 metres.

$$\text{Volume} = \frac{3.142 \times 0.5 \times 2}{4} = 0.39\text{m}^3$$

The volume of a liquid or gas is found by the same method.

Density

Density is the measure of how compact a substance is. If we compare two blocks of the same size, one made of iron and the other wood, the iron block would be the heaviest, so we can say that iron has a greater density than wood. The SI unit for density is kg/m^3 (kilograms per cubic metre). Its symbol is ρ (rho) and the sub multiples are kg/dm^3 (kilograms per cubic decimetre) and g/cm^3 (grams per cubic centimetre).

Mass

The mass of a body is a measure of the quantity of matter in the body and the SI Unit of mass (m) is the kilogram (kg).

Try this

1 Calculate the volume, in m^3 of a tank 3 metres by 2 metres by 1.5 metres.

_____

_____

2 Calculate the capacity, in m^3, of a cylindrical tank 0.5m in diameter and 1.5m high.

_____

_____

Try this

State the SI unit, the unit symbol and the variable symbol for the following. The first one is done for you.

Variable	Unit	Unit symbol	Variable symbol
length	metre	m	*l*
mass			
area			
velocity			
temperature			
density			

Part 3 Electrical variables

Electrical variables are also known as electrical quantities and they are covered in more detail in subsequent chapters.

Resistance (symbol R)

All materials at normal temperatures oppose the flow of current through them. This opposition to the current flow is the resistance (R) of the material. The SI unit of resistance is the ohm (Ω).

Good conducting materials, such as copper and silver, offer very little resistance under normal circumstances. There are, however, three factors that can change this:

● the length of the conductor
● the temperature of the conductor
● the cross-sectional area of the conductor.

The first factor is the length of the conductor. The longer the conductor the further the current has to flow, so the resistance is greater. For example 100m of a single cable is found to have a resistance of 1Ω. This means that 200m of the same cable will have a resistance of 2Ω.

The second factor is temperature. The higher the temperature the more resistance there is to current flow. Similarly, the lower the temperature the less resistance there is and

the greater the conductivity. Carbon is an exception to this, as it has a negative (opposite) temperature reaction.

The third factor is the cross-sectional area of the material. A thin conductor will have a greater resistance to current flow than a thicker conductor. As many of the conductors we use are round, and the measurements we can take from them are the diameters, care must be taken when working out calculations. Remember that the formula for working out the csa is

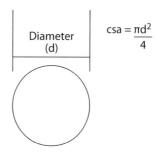

Figure 2.6 *Cross-sectional area of a circle*

So if you double the diameter of a conductor you increase its area by a factor of 4. This means that a conductor with a resistance of 1Ω is twice the diameter of one with 4Ω, assuming they are of the same material and length.

This can get very interesting if we alter both diameter and length.

Remember

Resistance is measured in ohms.

The SI unit symbol is Ω.

The variable or quantity symbol for resistance is R.

Resistance changes with conductor length, cross-sectional area and temperature change.

Example: Let us take an example of a round conductor with a diameter of 1mm and a length of 1 metre. We require another conductor of the same material with the same resistance, but not the same dimensions.

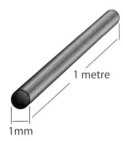

1 metre

1mm

Figure 2.7 *Round conductor*

If we first increase the diameter to 2 mm this means the cross-sectional area has increased by a factor of 4, so if we still had 1 metre in length the overall resistance would be less, in fact only $^1/_4$ of what it was before. To maintain the same resistance the length will need to be increased to 4 metres.

To calculate the resistance there are two formulae which take into account the four factors that affect resistance:

- material
- length
- temperature
- cross-sectional area.

$$\text{Resistance} = \frac{\text{resistivity} \times \text{length(l)}}{\text{cross} - \text{sectional area(A)}}$$

'Material', in this calculation, is the variable for resistivity, variable symbol ρ (rho) and SI unit symbol Ωm, so the formula can be written

$$R = \frac{\rho l}{A}$$

The resistivity of a material is the resistance of a sample of unit length and unit cross-sectional area. Material that is a good conductor has a low value of resistivity.

Material that is a poorer conductor has a higher resistivity.

Some examples of resistivity include

- copper has a resistivity of 17.2×10^{-9} Ωm
- aluminium has a resistivity of 28.4×10^{-9} Ωm
- nichrome has a resistivity of 1110×10^{-9} Ωm.

These are the values at a temperature of 20°C.

So the resistance of a piece of copper 1 metre long and 1 square metre in cross-sectional area is 17.2×10^{-9} Ω (or 0.000 000 017 Ω).

The second formula relates to the resistance after a temperature rise.

$$\frac{R_1}{R_2} = \frac{1 + a\, t_1}{1 + a\, t_2}$$

where

R_1 is the resistance at the start temperature
R_2 is the resistance at the new temperature
α is the temperature coefficient of resistance
t_1 is the start temperature (°C) and
t_2 is the new temperature (°C).

Try this

A conductor has a diameter of 2mm and a length of 2 metres. Determine the diameter of a similar conductor 8 metres in length if the two conductors are to have the same resistance.

_____

_____

The temperature coefficient of resistance(α) of a material is:

The increase in resistance of a material having a resistance of 1Ω at 0°C when its temperature is raised by 1°C.

Remember

The SI unit of resistivity is the ohm metre (Ωm).

Try this

Determine the resistance of 200m of 120mm² single-core copper cable where the resistivity of copper is 17.2×10^{-9} Ω m. Remember to show all working.

_____

_____

_____

A coil has a resistance of 80Ω at 0°C. Determine the resistance of the coil at 25°C. The temperature coefficient of resistance for the coil is 0.004 Ω/Ω °C at 0°C.

_____

_____

_____

Part 4

Voltage (symbol V)

To create the flow of electrons through a circuit pressure has to be applied. The force which moves the electrons is called the electromotive force (emf). The circuit pressure is measured in volts (V) and emf is usually applied at the source of a circuit. Like pressure in any system, there are parts where the pressure drops. This is referred to in an electrical circuit as a potential difference (pd), but it is still measured in volts (V).

Remember

Electromotive force (emf) is measured in volts.

The SI unit symbol is V.

Potential difference (pd) is also measured in volts (V).

Current (symbol I)

Electric current is a measure of the rate of flow of electrons through a conductor. Its SI unit is the ampere (A).

Now let us see what the relationship is.

The quantity of electricity is measured in coulombs (C) and one coulomb is equal to 6.3×10^{18} electrons, that is

1 coulomb = 6 300 000 000 000 000 000 electrons

One ampere of electricity is said to flow with the use of one coulomb of electricity in one second.

That is to say that one ampere of electricity flows each time 6.3×10^{18} electrons flow past a point in one second.

This can be shown as

$$\text{amperes} = \frac{\text{quantity of electricity in coulombs}}{\text{time in seconds}}$$

or $\quad I = \dfrac{Q}{t}$

or $\quad Q = It$

Power (symbol P)

Power is the rate at which energy is converted from one form to another, for example, electrical energy may be converted into heat energy, and is defined as the energy converted divided by the time taken for the conversion, hence:

$$\text{Power} = \frac{\text{Energy}}{\text{Time}}$$

The SI unit of power is the watt (W), energy is measured in joules (J) and time as we know is in seconds (s).

Therefore: $\text{Power in Watts} = \dfrac{\text{joules}}{\text{seconds}}$

When energy = 1 joule, and time = 1 second then power is 1 watt.

Therefore: 1 watt = 1 joule/ second or 1W = 1 J/s

i.e. One watt is the power when one joule of energy is converted in one second.

Example: Calculate the energy converted into heat by a 2kW electric heater in 1 hour.

Remember from before: $\text{Power} = \dfrac{\text{energy}}{\text{time}}$ and so Energy = power × time

Where: P = 2kW = 2000W and t = 1 hour = 60 min × 60s = 3600 seconds

∴ Energy = 2000 × 3600 = 7 200 000 joules or 7.2×10^6 J

Energy (symbol W)

Energy is the ability to do work.

The SI unit of energy is the joule (J), and one joule of energy is consumed when one watt of power is absorbed for a time of one second.

Remember from before: Energy (W) = power (W) = (watts) × time (seconds)

The joule or watt-second is a very small unit of energy, therefore, for practical reasons, the kilowatt-hour (kWh) is used to measure the energy consumed by domestic and small commercial and industrial consumers.

Therefore now we have: Energy in kWh = power (kW) × time (h)

Try this

1 Calculate the resistance of 125m of 50mm^2 aluminium cable. Resistivity of aluminium 28.4×10^{-9}

_____

_____

_____

2 How many coulombs of electricity are used when 25 amperes flow for 10 minutes?

_____

_____

_____

Example: If a 2kW electric heater is switched on for 4 hours, the energy consumed is:

Energy = 2 × 4 = 8kWh

The kilowatt-hour is often referred to as a unit of electricity, and so 8kWh = 8 units of electricity consumed by the heater.

By comparison, the energy consumed in joules is:

Energy = 2000 × (4 × 60 × 60) = 28 800 000J

The relationship between joules and kilowatt-hours is:

$$1\text{kWh} = 1\,(\text{kW}) \times 1\,(\text{h}) = 1000(\text{W}) \times (60 \times 60)\,\text{s}$$
$$= 3\,600\,000\,\text{J or } 3.6\,\text{MJ}$$

So 1 kilowatt-hour = 3.6 million joules.

Frequency and periodic time

The time it takes to complete one full cycle (360°) of the ac supply is known as the periodic time (T).

The number of full cycles completed in one second is known as the frequency (f).

The supply frequency can be calculated using the formula:

$$f = \frac{1}{T}$$

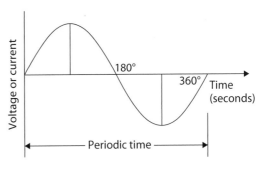

Figure 2.8 *Full cycle of an alternating voltage or current*

where f is frequency in hertz (Hz) and T is periodic time in seconds (s)

In the UK, the frequency of the supply is 50 cycles per second or 50Hz.

Example: What is the periodic time of a 50Hz ac supply?

$$T = \frac{1}{f} = \frac{1}{50} = 0.02\,\text{s or } 20\,\text{ms}$$

Remember

The SI unit for frequency is hertz (Hz)

Try this

Determine the frequency of the waveform shown.

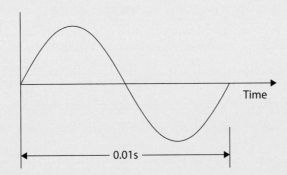

Figure 2.9 *Waveform*

Part 5 Inductance and inductive reactance

Figure 2.10 *An inductor with an iron core and the general symbol for inductor*

As an inductor is a coil of wire it has resistance, and when connected to a dc supply it will act as a resistor. However, when a coil is connected to an ac supply it becomes an inductor and takes on additional properties as it reacts with the alternating current. Inductance is measured in henrys, the variable symbol is L and the unit symbol is H. The alternating current creates the effect of a continuously changing magnetic field inside the coil. This effect reacts with the flow of current and opposes it. This means that when the coil is connected to a dc

supply it is only the resistance of the coil that limits the flow of current, but when connected to an ac supply there is also reactance due to the inductance of the coil. Inductive reactance, as it is called, is measured in ohms; the variable symbol is X_L and the unit symbol Ω.

Inductive reactance can be calculated using the formula:

$$X_L = 2\pi f L$$

Where X_L = inductive reactance (Ω), f = frequency of supply (H$_z$) and L = inductance (H).

Example: Calculate the inductive reactance of a coil which has a self inductance of 0.318H and negligible resistance when connected to a 230V, 50 Hz supply.

$$X_L = 2\pi f L$$
$$= 2 \times 3.142 \times 50 \times 0.318$$
$$= 100\,\Omega$$

The inductance of a coil can be increased by

● increasing the number of turns of wire on the coil
● increasing the iron core in the coil.

The capacitor

The capacitor is a device which can store an electric charge. A capacitor usually consists of two conductive plates, separated from each other by a layer of insulation known as a dielectric.

Try this

Determine the inductive reactance of a 1H non-resistive inductor when connected to a 230V 50Hz supply.

_____

_____

_____

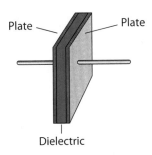

Figure 2.11 *The basic capacitor*

Capacitance (symbol C)

Capacitance is a measure of a capacitor's ability to store an electric charge and, put simply, a large capacitance means that more electric charge can be stored.

A capacitor which can store a charge of one coulomb at a potential difference of one volt is said to have a capacitance (C) of one farad (F)

$$C = \frac{Q}{V}\,(\text{Farads})$$

$$\text{or}\quad Q = CV\,(\text{Coulombs})$$

Example: A capacitor of 200μF which is fully charged at 200 volts holds a charge of

$$Q = CV$$

$$Q = 200 \times 10^{-6} \times 200$$

$$= 40\,\text{mC}$$

The value of capacitance

The value of capacitance is measured in farads (F) but because of the very large quantity of one farad, values are generally given in microfarads (μF). Occasionally nanofarads (nF) and picofarads (pF) are used for very small values.

$$1\mu\text{F} = 10^{-6}\,\text{F}$$

$$1\text{nF} = 10^{-9}\,\text{F}$$

$$1\text{pF} = 10^{-12}\,\text{F}$$

The capacitance of a capacitor is dependent on three main factors:

- the area of the facing plates: the larger the area the greater the capacitance
- the distance between the plates: the less the distance the greater the capacitance
- the type of the dielectricor spacing material.

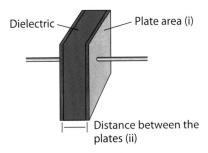

Figure 2.12 *Capacitor plates*

Some types of capacitor do not have solid dielectric material between the plates; they just have an air gap between the plates because air is quite effective as a dielectric. As can be seen from Table 2.4, however, there are dielectric materials which are better.

Try this

An 80μF capacitor is fully charged from a 230V supply. Calculate the charge stored.

Table 2.4 *Dielectric constants*

Material	Dielectric constant
Air	1.0
Aluminium oxide	10.0
Glass	7.6
Mica	7.5
Mylar	3.0
Paper	2.5
Porcelain	6.3
Quartz	5.0
Tantalum oxide	11.0

Capacitive reactance (X_C)

In an ac circuit the supply voltage and current are continually changing, so that the capacitor is either being charged or discharged more or less continually one way and then the other way for each half cycle of the ac supply. The opposition (offered by the capacitor) to the changing current in the ac circuit is called the capacitive reactance X_C. As with inductive reactance, capacitive reactance is measured in ohms (Ω).

Capacitive reactance can be calculated using the formula:

$$X_C = \frac{1}{2\pi f\, C}$$

Where X_C = capacitive reactance (Ω), f = frequency of supply (Hz) and C = capacitance (F).

However, if capacitance is in microfarads (μF) the formula is:

$$X_C = \frac{10^6}{2\pi f\, C}$$

Example: Calculate the capacitive reactance at 50Hz of a 47 μ F capacitor.

$$X_c = \frac{1}{2\pi fC} \text{ and so } X_c = \frac{10^6}{2 \times 3.142 \times 50 \times 47} = 67.72\Omega$$

Try this

Calculate the capacitive reactance at 50Hz of the following capacitors:

1   22 μ F

_____

_____

2   150 μ F

_____

_____

Part 6

Impedance (symbol Z)

The total opposition to the flow of alternating current in an ac circuit is known as the impedance (Z) of the circuit and its unit is the ohm (Ω). That is, impedance (Z) is the effective opposition to alternating current flow of all the components (resistance, inductance and capacitance) in the circuit.

One way to determine the impedance of an ac circuit is by drawing an impedance triangle and applying Pythagoras' theorem:

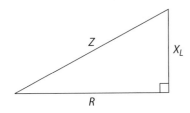

Figure 2.13 *Typical impedance triangle*

Where

X_L is inductive reactance in ohms

R is resistance in ohms

Z is impedance in ohms.

Since X_L and R are always shown at right angles Pythagoras' theorem can be used.

This means that:

$$Z^2 = R^2 + X_L^2 \text{ or}$$

$$Z = \sqrt{R^2 + X_L^2}$$

Example: A resistor of 30Ω is connected in series with an inductor of reactance 40Ω. Ignoring any resistance in the inductor, calculate the impedance of the circuit.

$$Z = \sqrt{R^2 + X_L^2}$$

$$= \sqrt{30^2 + 40^2}$$

$$= \sqrt{900 + 1600}$$

$$= \sqrt{2500}$$

$$= 50\,\Omega$$

The simplest right-angled triangle is a 3:4:5 triangle, i.e. if R is 3 Ω, and X_L is 4 Ω then Z must be 5 Ω.

a)   It can be seen if there is just resistance (R) and inductance (L) in the circuit, the formula is:

$$Z = \sqrt{R^2 + X_L^2}$$

b)   If there is just resistance (R) and capacitance (C) in the circuit, the formula is:

$$Z = \sqrt{R^2 + X_C^2}$$

c)   If there is resistance (R), inductance (L) and capacitance (C) in the circuit, and if X_L is greater than X_C, the formula is:

$$Z = \sqrt{R^2 + (X_L - X_C)^2}$$

Try this

Calculate the impedance of a circuit which has a 100Ω resistor connected in series with a capacitor of reactance 200Ω.

_____

_____

_____

and if X_C is greater than X_L the formula is:

$$Z = \sqrt{R^2 + (X_C - X_L)^2}$$

Remember

The SI unit of resistance (R), inductive reactance (X_L), capacitive reactance (X_C) and impedance (Z) is the ohm (Ω).

Try this

Determine the impedance of the ac series circuit shown below by applying Pythagoras' theorem.

$R=10\Omega$ $X_L=12\Omega$ $X_C=7.5\Omega$

_____

_____

_____

Power factor

The product of voltage and current gives the 'actual or true power' of a dc circuit (i.e. $P = V \times I$), and this is the amount of power actually consumed by the circuit. In an ac circuit the product of voltage and current ($V \times I$) is the 'apparent power' and this is only equal to the 'actual (true) power' if the voltage and current are in phase with each other, as in a purely resistive ac circuit.

The ratio of actual or true power to apparent power is known as the power factor. This can be found by constructing a 'power triangle' and applying trigonometry.

The inductive part of the circuit is called the 'wattless component' (VA_r), since it consumes no power. It only provides a magnetic field. The resistive part is called the 'wattful component' or 'actual (true) power' (W) since all the power is dissipated in the circuit resistance. The combination of the two is known as the 'apparent power' (VA).

By trigonometry:

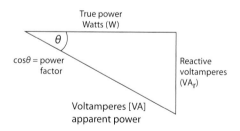

Figure 2.14 _Typical power triangle where power factor is lagging_

$$\cos \varnothing = \frac{\text{adjacent}}{\text{hypotenuse}}$$

$$\cos \theta = \frac{\text{true power}}{\text{apparent power}} = \frac{W}{VA}$$

$\therefore$ Power factor $=$ Cosine of the angle (θ)

Now power factor $= \cos\theta = \dfrac{kW}{kVA}$

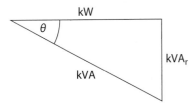

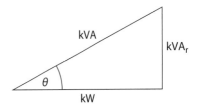

Figure 2.15 *The power triangle is usually shown in terms of kVA, kW and kVA$_r$.*

Figure 2.16 *Typical power triangle where power factor is leading*

Power factor can be one of three conditions:

- unity power factor
- lagging power factor
- leading power factor.

When the true power = apparent power, the power factor is unity (1), and the wattless power (kVA$_r$) is zero.

Power factors in inductive circuits are termed 'lagging' as the current lags the voltage.

Power factors in capacitive circuits are termed 'leading' as the current now leads the voltage.

Example: The actual power dissipated by the resistance in a single-phase ac inductive circuit is 60kW. Determine the power factor of this circuit if the apparent power of this circuit is: (a) 100kVA, (b) 80kVA,

a) $PF = \cos\theta = \dfrac{kW}{kVA} = \dfrac{60}{100} = 0.6$ lagging

b) $PF = \cos\theta = \dfrac{kW}{kVA} = \dfrac{60}{80} = 0.75$ lagging

Try this

Determine the power factor of a single-phase ac capacitive circuit if the true power dissipated by the resistance in this circuit is 20kW and the apparent power of this circuit is 22kVA.

_____

_____

Part 7

Power factor can also be calculated by applying trigonometry to the impedance triangle (as long as the values of resistance (R) and impedance (Z) are known).

Now $PF = \cos\theta = \dfrac{adj}{hyp} = \dfrac{R}{Z}$

It can clearly be seen from the calculations above that power factor does not have any units, it is just a ratio between the actual and apparent power of the circuit (or a ratio between the resistance and impedance of the circuit).

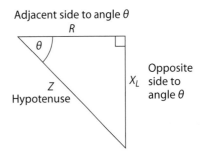

Figure 2.17 *Impedance triangle*

Try this:

Calculate the power factor of a single-phase ac inductive circuit with a resistance 45Ω and impedance 50Ω.

Power factor value

Power factor can be any value between 0 and 1 (unity), however, when the pf value is close to 1 (unity), the pf is good, and when the pf value is close to 0, the pf is very poor. Examples: 0.9 is a good pf and 0.4 is a very poor pf value.

Note: When Z = R the power factor of the circuit is unity and when the apparent power = actual power (kVA = kW) the power factor of the circuit is unity.

Power in ac circuits

We have already seen that in an ac circuit there are three types of power:

1 Actual or true power – the power dissipated in the circuit resistance and it is measured in watts (W) or kilowatts (kW).

2 Reactive power – the wattless power in the circuit and it is measured in volt-amperes reactive (VAr) or kilovolt-amperes reactive (kVAr).

3 Apparent power – the combined effects of the resistive and reactive power in the circuit and it is measured in volt-amperes (VA) or kilovolt-amperes (kVA).

We can determine the power in an ac circuit by applying Pythagoras' theorem to our power triangle.

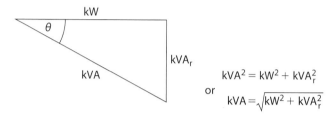

Figure 2.18 *Typical power triangle*

$$kVA^2 = kW^2 + kVA_r^2$$
$$\text{or} \quad kVA = \sqrt{kW^2 + kVA_r^2}$$

Example: Determine: (a) the kVA and (b) the power factor of a single-phase ac inductive load rated at 6kW and 8kVA$_r$.

a) $kVA = \sqrt{kW^2 + kVA_r{}^2} = \sqrt{6^2 + 8^2} = 10kVA$

b) $PF = \cos\theta = \dfrac{kW}{kVA} = \dfrac{6}{10} = 0.6 \text{ lagging}$

We can also determine power in an ac circuit by using trigonometry.

Example: Determine the kVA rating of a 15kW single-phase ac load with a power factor of 0.75 lagging.

Remember

$$PF = \cos\theta$$

$$\cos\theta = \frac{kW}{kVA} \quad \therefore \quad kVA = \frac{kW}{\cos\theta} = \frac{15}{0.75} = 20\,kVA$$

Try this

A single-phase ac capacitive load rated at 50VA and 45W. Determine the

1 VAr

2 power factor.

Try this

Determine the actual power rating of a 60kVA ac load with a pf of 0.85.

_____

_____

Part 8 Measuring instruments

When using test instruments take care not to use voltages and currents in excess of the component's rated value or the component may be destroyed. Before using the measuring instrument, check that it has the standard certification. This confirms that the calibration is correct.

Measuring resistance

The instrument used to measure resistance is an ohmmeter. As the values of resistance can range from a few ohms up to millions of ohms, the ohmmeter must also have ranges capable of measuring this. Traditionally, multi-range instruments have been used, where the ranges are switched as required. However self-ranging digital instruments are now most common.

Figure 2.19 _Measuring resistance with digital multi-meter_

Before measuring resistance it is important to first check the condition of the meter's internal battery. If this is low it can lead to inaccurate readings. Secondly, check that the meter has been set to zero with the leads connected together and switched to that required range.

Analogue readings

Measuring resistance with a meter which has an analogue scale can sometimes be confusing. The scale is not linear, and the zero is at the opposite end of the meter to the voltage and current scales.

© Megger

Figure 2.20 _A typical analogue multimeter_

Measuring voltage

Whenever it is necessary to take voltage readings care must be taken, as the readings have to be carried out when the circuit is live. To ensure that it is not possible to get a shock, safety precautions must be taken. If the voltage on the equipment you are working on exceeds 50V ac or 120V dc, special test probes must be used.

All test probes should meet the requirements of the Health and Safety Executive Guidance Note GS38.

Voltage can be measured across a component or between any two connections in a circuit. As with all testing it is important to have some idea what the reading should be. If this is not possible the meter should be switched to a high range and then brought down to a suitable range when you can see which range is appropriate.

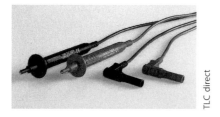

Figure 2.21 *Fused test prods complying with GS 38*

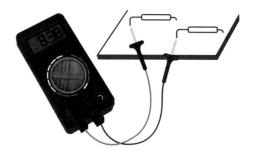

Figure 2.23 *Measuring current*

Clamp-on ammeter (tong tester)

The clamp-on ammeter allows current measurements to be taken without having to disconnect the supply and the circuit.

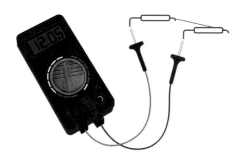

Figure 2.22 *Measuring voltage*

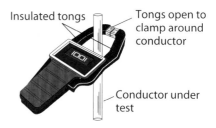

Figure 2.24 *Digital clamp-on ammeter*

Remember

The instrument used for measuring voltage may be the same multi-meter as used for resistance but switched to the voltage range.

Remember

This type of meter should be used on single insulated conductors where there are no exposed live parts. Use extreme caution when working around bare conductors or bus bars. Accidental contact with the conductor could result in electric shock.

Safety

Use the correct test probes which are designed for your safety.

Measuring current

When an instrument is connected to measure current it should be in series with the load, so that the full current flows through the meter. It is important that the instrument is suitable for the current that is being measured.

The supply should be switched off when the meter is connected and disconnected. Often a circuit has to be broken to connect the meter in series with the load. The circuit should be reconnected after the readings have been completed.

Analogue type clamp meters are available, but the digital types are most common, nowadays and the majority of these are of the clamp multi-meter type. They have the same functions/ranges as a digital multi-meter and some will measure the frequency of the supply.

Ammeter and voltmeter method for impedance measurement

In an ac circuit an ammeter and voltmeter having suitable ranges can be used for the determination of impedance (Z).

Applying Ohm's Law, with impedance in place of resistance (Z in place of R):

$$\text{Impedance} = \frac{\text{Voltage}}{\text{current}} \text{ or } Z = \frac{V}{I} \text{ and so Impedance}$$

$$= \frac{\text{Voltmeter reading (V)}}{\text{Ammeter reading (A)}}$$

Example: An ammeter and voltmeter are used to determine the impedance of an iron-cored inductor connected to a single-phase 50Hz ac supply. Neglecting iron loss and loading effects, calculate the impedance of the inductor if the voltmeter reading is 220V and the ammeter reading is 2A.

$$Z = \frac{V}{I} = \frac{220}{2} = 110\,\Omega$$

Remember

Voltage and current measurements are taken on live circuits. Every safety precaution must be used.

Try this

Calculate the resistance and impedance of an iron-cored inductor from the meter readings given below. Neglect iron loss and loading effects.

	Voltmeter readings	Ammeter readings
a) dc supply	12V	2.4A
b) ac supply	195V	15A

_____

_____

_____

Part 9

Analogue instruments

There are two basic types of analogue instrument, the moving coil and moving iron.

In the majority of cases an analogue instrument takes an electric current and uses this to produce a mechanical deflection of a pointer across a scale.

Moving coil

A coil of wire suspended between the poles of an electromagnet produces a force of deflection which is controlled by hair springs. The direction of the deflecting force depends on the direction of current flow in the coil and therefore this instrument cannot respond to an alternating current. This type of instrument is primarily intended for use in dc circuits only. With the aid of rectification, the moving coil instrument can be used for ac measurement provided that certain modifications are carried out.

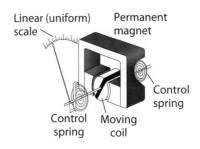

Figure 2.25 *Construction of a moving coil meter*

The scale of the moving coil instrument is linear, i.e. the scale divisions are the same size from zero to full-scale deflection (fsd.).

Moving iron

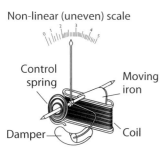

Figure 2.26 *Construction of an attraction type moving iron meter*

Moving iron instruments work on the principle of magnetic attraction or repulsion depending on their design. They normally use a coil spring controlling mechanism but some may use a simple gravity device. The moving iron instrument is deflected by direct current or rms alternating current and therefore it does not need a rectifier when connected to ac circuits.

The scale of a moving iron instrument is non-linear (uneven).

Analogue instruments are generally no longer used for testing electrical installations. Electronically operated digital instruments have replaced the analogue type instruments; however, analogue instruments are still used as measuring instruments in other areas of electrical work.

Digital instruments

Figure 2.27 *Typical digital multi-meter*

Digital instruments fulfil the same function as analogue instruments in that they are used to take measurements and to convey the information to the user.

The operating principle is, however, quite different to the analogue types previously discussed.

The most obvious difference is the display. This is generally an alpha-numeric arrangement with seven elements (segments) although more sophisticated types are readily available. This digital display is usually a liquid crystal display (LCD).

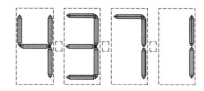

Figure 2.28 *Typical digital display*

A digital instrument also requires a power source before it can be operated. This will be a battery or batteries, in the case of the portable instrument, or a mains power supply in the case of bench or panel instruments.

The biggest difference is the operation of the digital instrument. The measured signal will be a voltage or, in the case of current or resistance measurement, it will be the voltage drop across a known value of resistor.

Digital instruments are generally more accurate than analogue instruments and the digital display is usually clearer to read than on an analogue type.

Extending the range of ammeters and voltmeters

In an ammeter, the basic movement is designed to give full-scale deflection at a very low current, far lower than would be considered suitable for practical applications. To extend the range so that the instrument can be of some practical use, a shunt resistor is connected in parallel with the meter movement. These shunts are 'low ohmic' value resistors.

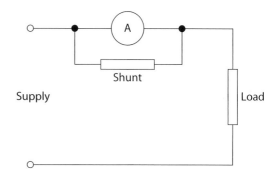

Figure 2.29 *An ammeter shunt*

Voltmeters

In the case of the voltmeter, the range is extended by connecting additional resistance in series with the meter movement. This series resistance is known as a multiplier and these are 'high ohmic' value resistors.

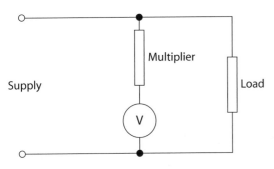

Figure 2.30 *Voltmeter multiplier*

Remember

The 'shunt' extends the range of a moving coil ammeter to read values of current higher than the instrument's movement is designed for.

The 'multiplier' is used to extend the range of a moving coil voltmeter.

Try this

The two basic types of analogue instrument are

a) _____

b) _____

A rectifier is fitted within a _____ meter to enable it to measure ac.

The type of instrument which has a non-linear scale is the _____

The type of instrument which has a linear scale is the _____

A _____ instrument requires a battery or a _____ supply for it to operate.

A shunt is used to extend the range of an _____

A multiplier is used to _____ the range of a _____ .

Part 10 Measurement of power

The power in a simple dc circuit can be measured using a voltmeter and an ammeter.

The readings of each are multiplied together to give the power reading. As power = voltage × current (P = V × I),

this can be carried out in an instrument which measures both voltage and current and displays the result in watts (a wattmeter). True power of an ac circuit can be measured with an electrodynamic (dynamometer) wattmeter.

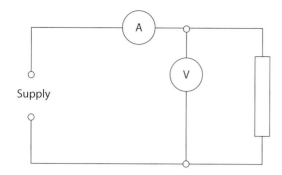

Figure 2.31 *Voltage and current measurement for VA*

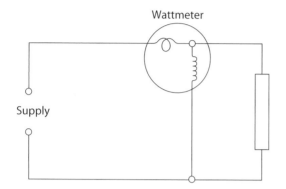

Figure 2.32 *Wattmeter circuit*

The current coil is connected in series, the same as a separate ammeter, and the voltage coil in parallel, the same as a voltmeter.

Measurement of energy

Energy is the measurement of power taken over a period of time.

It is an energy meter that the electricity suppliers use to monitor the electricity used by a consumer. The meter is often referred to as the kilowatt-hour meter as these are the quantities that are measured.

The energy, or kilowatt-hour meter, is connected in a similar way to the wattmeter because it also has current and voltage coils.

The older type energy meter had the voltage and current coils arranged in such a way that the magnetic fields being produced by them induced a current into an aluminium disc to make it act as a motor. The revolutions of the disc directly relate to the amount of energy consumed. Modern energy meters use electronic pulses to measure the energy used. The arrangements of the connections are the same.

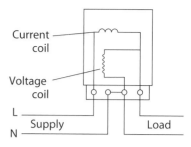

Figure 2.33 *Energy meter circuit*

Where there are large loads in use the current is monitored using current transformers around the main supply conductors.

Measuring frequency

Digital multi-meters and clamp multi-meters have functions for measuring frequency and are widely used nowadays.

Power factor measurement

Power factor can be measured by: (a) a power factor meter.

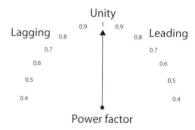

Figure 2.34 *Typical scale of a power factor meter*

Power factor meters are connected into circuits in a similar way to wattmeters.

(b) using a wattmeter, voltmeter and ammeter connected as shown in Figure 2.35.

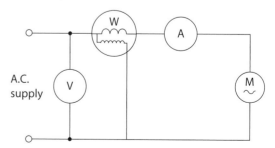

Figure 2.35 *Power factor measurement using a wattmeter, voltmeter and ammeter*

The wattmeter measures the 'actual or true power', and the voltmeter and ammeter arrangement measures the 'apparent power'.

$$\text{power factor} = \frac{\text{true power (wattmeter reading)}}{\text{voltage} \times \text{current}}$$

The answer would normally be less than unity (one).

This is a simple but cumbersome method of measuring power factor and it gives no indication whether the power factor is lagging or leading like a purpose made power factor instrument does.

Try this

When connected into a motor circuit a wattmeter reads 5kW, a voltmeter reads 230V and the ammeter reads 32A. Calculate the power factor of the motor.

_____

Test probes should conform to Health and Safety Executive Guidance Note _____ .

A clamp-on ammeter allows you to measure the _____ taken by a load without having to _____ the supply and the _____ .

A wattmeter measures the _____ power of the circuit and a voltmeter and ammeter measure the _____ power of the circuit.

Energy meters measure the amount of _____ taken over a period of _____ in _____ .

Congratulations you have now completed Chapter 2 of this study book. Correctly complete the self-assessment questions before you progress to Chapter 3.

SELF ASSESSMENT

Circle the correct answers.

1   Velocity is measured in the unit:

   a.   metre/second^2

   b.   square metre

   c.   metres per second

   d.   cubic metre.

2   The cross-sectional area of trunking whose dimensions are 75mm × 75mm is:

   a.   5625m^2

   b.   56.25m^2

   c.   562.5mm^2

   d.   5625mm^2

3   A conductor has a diameter of 4mm and a length of 2 metres. A similar conductor has a diameter of 2mm. Calculate the length of the second conductor if the resistances of the two conductors are to be the same.

   a.   0.5m

   b.   1m

   c.   8m

   d.   4m.

4   (i)   Resistance measurements are always taken with the supply to the circuits disconnected.

   (ii)   The ohmmeter is powered from its own internal supply.

Considering the statements (i) and (ii) above.

   a.   only statement (i) is correct

   b.   only statement (ii) is correct

   c.   both statements are correct

   d.   neither statement is correct.

5   In an ac circuit the wattmeter indicates the:

   a.   reactive power

   b.   actual power

   c.   apparent power

   d.   power factor.

3 Mechanical science

RECAP

Before you start work on this chapter, complete the exercise below to ensure that you remember what you learned earlier.

The farad is the SI unit for _____.

The hertz is the SI unit for _____.

The ohm is the SI unit for _____.

The unit for inductive and capacitive reactance is the _____.

The unit for energy is the _____ .

The unit for power is the _____ .

Power factor is the ratio between the actual power and the _____ power.

An ohmmeter is used to measure _____.

A wattmeter measures _____.

LEARNING OBJECTIVES

On completion of this chapter you should be able to:

● Specify what is meant by mass, force and weight.

● Explain the basic mechanical principles of levers, pulleys and gears.

● Describe the main principles and interrelationships of force, work, energy, power and efficiency.

● Calculate values of electrical energy, power and efficiency.

Part 1 Mass, force and weight

Mass is the quantity of matter that a body contains and it is measured in kilograms. The SI unit symbol is kg and the quantity symbol *m*. A body's mass can be found by weighing the body and comparing it with known standard masses.

5kg 2kg 1kg 500g 200g 100g

Figure 3.1 *Measuring a body using standard masses*

Force, at this stage, is best considered as an influence which tends to cause a body to move. The SI unit of force is the newton, symbol N.

If an object is stationary, or moving in the direction of an applied force, it will increase speed in the direction of the applied force. This is known as acceleration and the greater the applied force the greater the acceleration.

Figure 3.2 *Accelerating a football*

If the object is already moving in the opposite direction to the applied force, the effect of the force is to slow the body down, known as deceleration.

An object left on level ground remains stationary. This is because the earth exerts a force on all masses. We know this force as gravity. The earth's 'gravitational field' varies only slightly at any point on the earth's surface and it is generally accepted that a mass of 1kg at the earth's surface experiences a force of 9.81 newtons due to gravity.

Weight can be considered as the product of the mass of an object and the force of the earth's gravitational force.

Inertia is the name given to the property of an object continuing in an existing state of either rest or constant motion.

The force of an object may be considered as the product of the mass of the object and its acceleration. This can be expressed as

$$F = m \times a$$

where *F* is the weight of the body in newtons and *m* is the mass in kg and *a* is the acceleration in m/s^2 (metres per second squared).

An object falling freely in the earth's gravitational field is taken to have an acceleration of 9.81 m/s^2, so:

$$\text{newtons} = \text{mass in kilograms} \times 9.81 \text{ m/s}^2$$

The kilogram is the unit of mass and therefore cannot be used as a unit of force. The newton is the unit of force and we can consider that an object, with a mass in kg, will require a force in newtons to raise it against the force of gravity. It therefore follows that the object must exert this same force on any surface upon which it is placed.

The area of the object, along with the force exerted upon it exerts **pressure** on the surface upon which it is placed. This pressure is a product of the force and area and is therefore measured in N/m^2.

Example:

A mass of 400kg has a force of 50N applied to it. What is the acceleration?

$$F = m \times a$$

Rearranging the formula:

$$a = \frac{F}{m}$$

Substituting the known quantities:

$$a = \frac{50}{400}$$
$$a = 0.125$$

The acceleration is 0.125 m/s^2.

Example:

Some trunking has a mass of 400kg. What force will be required to lift it?

$$F = m \times a$$
$$F = 400 \times 9.81$$
$$= 3924.00 \text{ N}$$

Try this

1 A mass of 300kg has a force of 30N applied to it. What is the acceleration?

_____

2 A mass of 80kg is placed on a table. Calculate the downward force on the table.

_____

3 What force would be required to lift a motor having a mass of 350kg?

_____

Remember

Mass is the quantity of matter that a body contains.

Force is an influence tending to cause the motion of a body.

Weight is the force experienced by a body due to the earth's gravity.

A force has both magnitude and direction. We can show this as a line representing the force applied to a body. The length of the line, drawn to a suitable scale, represents the magnitude, or size, of the force and the direction of the force is shown by the direction of the line.

Figure 3.3 *Basic vector diagram*

When using vectors to represent force on a body it is assumed that the force acts upon a single point, and the direction of the force is illustrated by the angle of the line with an arrow placed at the end of the line indicating the direction in which the force operates.

These vector diagrams are used in electrical calculations where they relate to electrical quantities and are referred to as phasor diagrams.

If two forces act upon a body in the same direction the resultant will be the sum of the forces:

$$6\,N + 8\,N = 14\,N$$

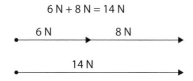

Figure 3.4 *Basic vector addition*

If these two forces act in opposite directions the result will again be the mathematical sum of the forces.

$$8\,N - 6\,N = 2\,N$$

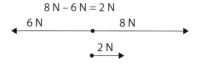

Figure 3.5 *Basic vector subtraction*

When the two forces are opposite and equal then the body is said to be in **equilibrium**, and there would be no resultant direction.

In reality, forces are often acting at an angle to one another. In such cases we often need to find the resultant force. We can do this by producing a parallelogram of the forces. Where two forces are operating at 90° to one another for example:

Figure 3.6 *More complex vector diagram*

Completing the parallelogram of forces we have:

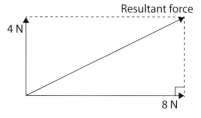

Figure 3.7 *Complex vector addition using parallelogram of forces*

The resultant force is the diagonal of the parallelogram giving both the magnitude and the direction of the force. If the vector is drawn to scale the value of the resultant force may be measured, as may be the angle at which the force operates.

We can calculate the magnitude of the force using Pythagoras' theorem for right-angled triangles.

$$\text{resultant force} = \sqrt{\text{force 1}^2 + \text{force 2}^2}$$
$$= \sqrt{8^2 + 4^2}$$
$$= \sqrt{64 + 16}$$
$$= \sqrt{80}$$
$$= 8.94\,\text{N}$$

We can apply the parallelogram of forces irrespective of the angle at which the forces operate providing we remember the basic rules of keeping the angle and length the same for the parallel sides. For example:

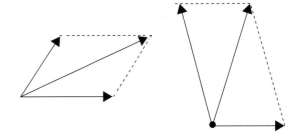

Figure 3.8 *Vector addition using parallelogram of forces*

Try this

Use a scale of 1cm representing 1N and complete the parallelogram for the following forces. Draw in the diagonal and measure the resultant force.

1   Forces of 4N and 3N at an angle of 60° to each other.

_____

2   Forces of 2.5N and 5N at an angle of 90° to each other.

_____

Define the following:

Mass is _____

_____

Force is _____

_____

Weight can be considered as _____

_____

Force is a vector quantity. This means it has both magnitude and _____ .

Part 2 Levers, pulleys and gears

Levers

Levers, in the form of crowbars, are used in order to raise a heavy load with a small effort. The same principle applied to claw hammers.

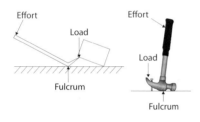

Figure 3.9 *Basic first-order levers*

A lever is a bar pivoted so as to be able to rotate about a point. This point is called the fulcrum.

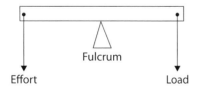

Figure 3.10 *First-order lever*

The position of the fulcrum is important. The nearer it is to the load the less force has to be applied in order to lift the load.

There are three common arrangements for simple levers. The crowbar is a first-order (or class 1) lever and the fulcrum is between the load and the effort. Second-order levers are where the fulcrum is at one end and the load is nearest to the fulcrum. A wheelbarrow is an example of a second-order lever. In a third-order lever arrangement the load is at the opposite end of the lever to the fulcrum, and in this case the effort is always greater than the load. Tweezers or woolshears are examples of third-order levers.

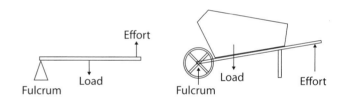

Figure 3.11 *Second-order levers*

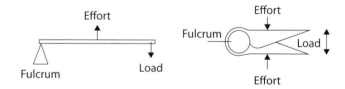

Figure 3.12 *Third-order levers*

Moments

The turning effect of the force is called the moment of the force. A moment is measured in newton metres (Nm).

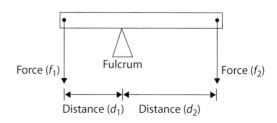

Figure 3.13 *First-order terms*

moment = force × distance

$$= f_1 d_1 \text{ or } f_2 d_2$$

A moment can cause a clockwise or an anticlockwise turn around the pivot or fulcrum.

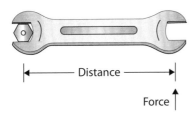

Force

Figure 3.14 *An anticlockwise moment*

A state of **equilibrium** is said to exist if the anticlockwise moment equals the clockwise moment and the system is at the point of balance. This is called the **principle of moments**. In Figure 3.15 the downward forces are F_1 and F_2. The distances between the fulcrum and the downward forces are d_1 and d_2. In a state of equilibrium:

$$f_1 d_1 = f_2 d_2$$

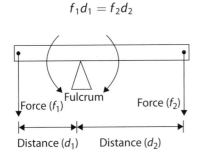

Force (f_1)   Fulcrum   Force (f_2)

Distance (d_1)   Distance (d_2)

Figure 3.15

Example:

What effort would be needed to lift a load of 25kg placed 200mm from the fulcrum of the lever? The distance

between the fulcrum and the point the force is to be exerted is 1 metre. Remember the force exerted by the 25kg load is 25 × 9.81 newtons.

$$f_1 d_1 = f_2 d_2$$

$$f_1 = \frac{f_2 d_2}{d_1}$$

$$= \frac{25 \times 9.81 \times 0.2}{1}$$

$$= 49.05 \text{ N}$$

Remember

A moment is the turning effect on an object produced by a force acting at a distance.

A lever is a rigid bar pivoted about a fulcrum which can be acted upon by a force in order to move a load.

A fulcrum is the point against which a lever is placed.

Pulleys

A pulley is a mechanism composed of a wheel on an axle or shaft that usually has a grooved rim around its circumference. A rope, belt, cable or chain runs over the wheel and inside the grooved rim.

Types of pulley system

Fixed. A fixed (or class 1) pulley has a fixed axle and is used to change the direction of the force on a rope.

Try this

What effort would be needed to lift a load of 25kg placed 500mm from the fulcrum of a lever? The distance between the fulcrum and the point the force is to be exerted is 900mm.

_____

_____

_____

A fixed pulley has a mechanical advantage of 1 (the force is equal on both sides of the pulley, so there is no multiplication of force).

Movable. A movable (or class 2) pulley has a free axle and is used to multiply forces. A movable pulley has a mechanical advantage of 2 (if one end of the rope is anchored, pulling on the other end of the rope will apply double the force to the object attached to the pulley).

Compound. A compound pulley is a combination of a fixed and a movable pulley system.

Pulleys are used to change the direction of an applied force, transmit rotational motion or provide a mechanical advantage.

The pulley block

The pulley block consists of a continuous chain or rope passing over a number of pulley wheels. Pulley blocks should be regularly tested and their safe working load displayed on them, which should *never* be exceeded.

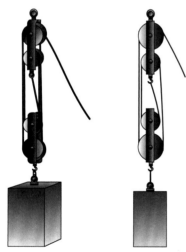

Figure 3.16 *A four-pulley system*

When loads are suspended on pulley systems they have a tendency to swing and twist. This problem is often overcome by having a stabilizing or control rope tied to the load and by one person being given the sole responsibility of keeping the load straight.

Loads should never be left suspended in mid-air without someone to watch them. The area under the load should be kept clear at all times in case the load should fall.

When the load has been lowered gently into position it should be checked for movement before the sling is taken away.

Why use pulleys?

Pulleys can give several advantages when trying to lift an object. If a single wheel pulley is used there is no mechanical advantage but a difficult shaped object can be slung so that a single rope can lift it.

Where a pulley system with two pulleys is used the amount of effort required is only half that of a one-pulley system. If we go to a four-pulley system, the effort required is only a quarter of that required with a single pulley.

Example:

When lifting a load of mass 16kg with a single pulley the effort required is also 16kg.

Using a two pulley system the same load can be lifted with an effort of:

two pulleys require half the effort, i.e. 8kg

With four pulleys the same load can be lifted with an effort of:

four pulleys require one quarter the effort, i.e. 4kg.

Try this

1   What mass can be lifted with an effort of 15kg on a two-pulley system?

_____

2   On a four-pulley system a load of 36kg has to be lifted. What effort is required?

_____

3   A man has to lift a mass of 72kg. The effort he uses is 18kg. What pulley system does he require in order to lift the load?

_____

Part 3

Belt drives

Belt drives are often used to transmit rotational motion from the shaft of one machine to the shaft of another machine. The shaft on each machine will have a pulley wheel fixed to it and one or more belts over the two pulleys.

The majority of belts used for connecting an electric motor to a machine are of the 'vee' type, which grip the pulley better than a flat belt and therefore reduce slip.

Gears

A gear is a rotating machine part with cut teeth, or cogs, which mesh with another toothed part in order to transmit torque (or rotational motion).

A gearbox or transmission using two or more gears working in tandem can produce a mechanical advantage through a gear ratio (for example the automobile transmission allows selection between gears to give various mechanical advantages).

Types of gear

There are numerous types of gear: external, internal, spur, worm and helical are a few examples. Spur gears or straight-cut gears are the simplest type of gear.

Figure 3.17 *(a) Helical gear, (b) Spur gears*

Gear drives are more positive than belt drives, as they cannot slip, but they may be noisy and will usually require periodic lubrication.

Gear sizes are usually indicated by the number of teeth rather than by diameter, the two being proportional.

Similar to pulleys, two gears of the same size enmeshed and turning together will have the same peripheral (outside edge) speeds and identical torques, if losses are neglected.

If the output shaft of a gearbox is designed to run slower than the input shaft, its output torque will increase and if the output shaft is designed to run faster than the input shaft, its output torque will decrease.

Gear ratio

Gears are normally used for one of four reasons:

1   To increase or decrease the speed of rotation.
2   To reverse the direction of rotation.
3   To change rotational motion to a different axis.
4   To keep the rotation of two axes synchronized.

The gear ratio is the relationship between the numbers of teeth on two gears that are meshed together.

Figure 3.18 *A 3:1 gear ratio*

If the diameter of one gear is three times that of the other gear the gear ratio is 3:1 (every time the larger gear goes around once, the smaller gear goes round three times).

The teeth on gears have three advantages:

1   They prevent slipping between gears.
2   They make it possible to determine exact gear ratios (if the larger gear has 80 teeth and the smaller gear has 20 teeth, the gear ratio is 4:1).
3   Any slight imperfections in the actual diameters of two gears do not matter, because the gear ratio is controlled by the number of teeth even if the diameters are not precise.

Remember

There are three types of pulley system: fixed, movable or compound.

Pulleys are used to provide a mechanical advantage, transmit rotational motion or change the direction of an applied force.

Vee-type belts are often used because they grip the pulley better than a flat-type belt.

Gears are rotating wheels with teeth that mesh together to transmit torque.

Gear drives are better than belt drives as they cannot slip.

Gears are normally used to increase or decrease speed, reverse direction of rotation, change rotational motion to a different axis or to keep the rotation of two axes synchronized.

Work, energy and power

Let's first look at the physical quantities in Table 3.1 before considering work, energy and power, so we can clearly see their relevance in later calculations.

Table 3.1 *Physical quantities*

Physical Variable	Unit	Definition
Force (*F*)	newton (N)	Force is a dynamic influence on a body or object which can cause a change in either the shape or the motion of the body or object.
Mass (m)	kilogram (kg)	Mass is the amount of matter which is contained in a body or object.
Weight	newton (N)	Weight is the downward force due to the mass of a body or object.

There is a connection between the mass of an object and its weight which is:

$$\text{Weight} = \text{mass} \times 9.81\text{N}$$

Remember that 9.81N is the force required to raise a mass or load of 1kg against the effect of gravity.

Example:

Calculate the weight of a bundle of steel conduit which has a mass of 100kg.

$$\text{Weight} = \text{mass} \times 9.81\text{N}$$
$$= 100 \times 9.81\text{N}$$
$$= 981\text{N}$$

Work

Work is done whenever an object is moved by applying a force to it, and is calculated from:

$$\text{Work done} = \text{force applied} \times \text{distance moved}$$

$$W = F \times d$$

Force is in newtons and distance is in metres.

Try this

Determine the weight of a transformer with a mass of 400kg.

_____

_____

The unit of work is the joule (J) which is defined as the amount of work done when a force of 1 newton acts through a distance of 1 metre.

$$1J = 1Nm$$

Example:

The work done in lifting a mass of 10kg through a height of 6m is:

$$\text{Weight} = \text{mass} \times 9.81N$$
$$= 10 \times 9.81N \qquad = 98.1N$$
$$W = F \times d$$
$$= 98.1N \times 6m$$
$$= 588.6 \text{ Newton metres (Nm)}$$
$$\text{OR } 588.6 \text{ joules (J)}$$

Try this

A hoist lifts a mass of 500kg through a vertical distance of 30m to the top of a building. Calculate the work done by the hoist in kilojoules.

_____

_____

_____

Part 4

Energy

The capacity to do work is called energy, which comes in many forms. Some of the most significant forms of energy are atomic energy, chemical energy, heat energy, mechanical energy and electrical energy. To use energy we often need to convert energy from one form to another. Take, for example, a coal-fired power station.

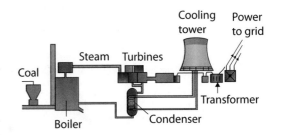

Figure 3.19 _Coal-fired power station_

Coal contains chemical energy which is converted, through combustion, into heat energy. This heat energy is used to produce steam, which in turn is the energy used to drive a steam turbine which converts the heat energy into mechanical energy. The turbine drives a generator which converts the mechanical energy into electrical energy.

During this conversion, energy can be neither created nor destroyed, so the total energy output is equal to the energy input. The useful energy output of a machine is always less than the input. This is due to the losses within the machine.

The unit of all forms of energy is also the joule (J), but it is often more convenient to measure electrical energy in kilowatt-hours (kWh).

$$3\,600\,000\text{J} = 1\text{kWh}$$

$$\text{from} \quad J = W \times t$$

$$= 1000\text{W} \times 60 \text{ min} \times 60 \text{ sec}$$

$$= 3\,600\,000\text{J}$$

Note

Energy and work are interchangeable since energy must be used to do work.

Both are measured in newton metres or joules.

So:

$$\text{Energy used (W)} = \text{Work done (W)}$$
$$= \text{Force applied (N)} \times \text{Distance moved (m)}$$

Example:

Calculate the energy required to raise a mass of 6kg through a vertical distance of 12m.

$$\text{Work done} = \text{force} \times \text{distance}$$
$$= \text{mass} \times 9.81 \times \text{distance}$$
$$= 6 \times 9.81 \times 12$$
$$= 706.32\text{Nm}$$
$$\text{or } 706.32\text{J}$$

Example:

A 250V dc generator provides a current of 10A. Calculate the energy dissipated (used) in 2 minutes.

$$P \qquad = U \times I$$
$$= 250 \times 10$$
$$= 2500 \text{ Watts}$$
$$\text{Energy} \quad = P \times t$$
$$= 2500 \times 120$$
$$= 300\,000 \text{ Joules (or 300kJ)}$$

Try this

An item of switchgear has a mass of 150kg.

Calculate:

1 the weight of the switchgear.

_____

_____

2 the energy used in raising the switchgear a distance of 3m above finished floor level.

_____

_____

Try this

Calculate the energy dissipated in 1 hour by a 250V, 20A dc generator. (Give the answer in megajoules.)

_____

_____

Kinetic energy (KE)

The kinetic energy of an object is the energy due to its motion.

The motion of the object can be in any direction, vertically, horizontally or at any angle.

There are numerous forms of kinetic energy including energy due to:

● rotational motion

● vibrational motion (oscillation)

● motion from one point (location) to another.

The kinetic energy of an object depends on the mass (m) of an object and the velocity or the speed (v) of the object.

Kinetic energy can be calculated by using different formulas (depending on the type of motion). We will use the following formula:

$$KE = \frac{1}{2}mv^2$$

where m is in kilograms (kg), v is in metres per second (m/s), KE is in joules (J).

Since the kinetic energy increases with the square of the velocity, an object doubling its velocity has four times as much kinetic energy.

Example:

Determine the kinetic energy of an object with a mass of 500kg and a velocity of 20m/s.

$$KE = \frac{1}{2}mv^2$$

$$KE = 0.5 \times 500 \times 20^2$$

$$= 100\,000J \text{ or } 100kJ.$$

Remember

Energy has the same units as work and work is force times distance ($W = F \times d$).

One joule is one newton of force acting through one metre ($1J = 1Nm$).

Example:

If Sid's car has a mass of 800kg and moves with a velocity of 25m/s, what force is required to stop the car in 40 metres?

Figure 3.20 _Sid's car_

$$KE = \frac{1}{2}mv^2$$

$$KE = 0.5 \times 800 \times 25^2$$

$$= 250\,000J$$

$$KE = F \times d \quad \therefore F = \frac{KE}{D} = \frac{250\,000}{40} = 6,250\,N \text{ or } 6.25\,kN$$

Try this

Calculate:

1   the kinetic energy of a motor vehicle with a mass of 850kg travelling at 18m/s.

_____

_____

2   the force required to stop the vehicle in 50 metres.

_____

_____

Part 5

Potential energy (PE)

Potential energy is stored energy that depends upon the relative position of various parts of a system.

Potential energy can be elastic or gravitational.

Elastic potential energy (EPE) is a measure of the returning force when an object changes its shape.

Most elastic objects will return to their original shape when the force is removed (for example, if we stretch a spring and let go, it will return to its original shape provided it has not been stretched beyond its elastic limit).

Gravitational potential energy (GPE) is a measure of how far an object can fall. The higher the object is, the further it can fall and the more GPE it will have. GPE also depends on the object's weight.

Gravitational potential energy (GPE) can be calculated by using the formula:

$$GPE = mgh$$

Where m is the mass of the object in kilograms (kg), g is the gravitational acceleration of the earth (9.81 m/s^2), h is the height above the earth's surface in metres (m) and GPE is in joules (J).

Example:

What is the gravitational potential energy of an object which has a mass of 50kg and is raised to a height of 5 metres?

$$GPE = mgh$$
$$= 50 \times 9.81 \times 5$$
$$= 2452.5 \text{ J}$$

Try this

An electrician accidentally drops a 1.36kg lump hammer from a mobile scaffold 4.5 metres above the ground. What is the hammer's potential energy.

_____

_____

_____

Efficiency

The efficiency of a machine (such as an electric motor) can be expressed as the ratio between the useful energy output and the total energy input, thus

$$\text{Efficiency } (\eta) = \frac{\text{useful energy output}}{\text{total energy input}} \times 100\%$$

The Greek letter η (termed eta) is the symbol for efficiency.

Example:

Calculate the energy in kWh required to lift a mass of 250kg through 200 metres if the efficiency of the hoist is 30%.

Figure 3.21 *Hoist*

$$\begin{aligned}\text{Useful work done} &= \text{force} \times \text{distance} \\ &= 250 \times 9.81\text{N} \times 200\text{m} \\ &= 490500\text{Nm (joules)}\end{aligned}$$

Transpose the formula for efficiency above:

$$\begin{aligned}\text{Total work done} &= \frac{\text{useful work done}}{\eta} \times 100 \\ &= \frac{490500}{30} \times 100 \\ &= 1635000 \text{ joules}\end{aligned}$$

As the energy is required in kWh:

$$\begin{aligned} &= \frac{1635000}{3600000} \\ &= 0.454 \text{ kWh}\end{aligned}$$

In most electrical and mechanical systems, it is more usual to express efficiency in terms of power rather than energy.

Power

Power (*P*) is a measure of how quickly work is done.

$$\begin{aligned} P \text{ (in watts)} &= \text{Rate of doing work} \\ &= \frac{\text{work done (in joules)}}{\text{time taken (in seconds)}} \end{aligned}$$

$$\begin{aligned} \text{or watts} &= \frac{\text{joules}}{\text{time}} \\ W &= \frac{J}{t} \end{aligned}$$

One watt is equivalent to work being done at the rate of one joule per second (1W = 1J/s).

$$\text{Energy used} = \text{work done}$$

$$\therefore \text{power (in watts)} = \frac{\text{energy (in joules)}}{\text{time (in seconds)}}$$

Transposed:

$$\text{Energy} = \text{power} \times \text{time.}$$

Example:

Calculate the power rating of an electric hoist motor required to raise a load of 200kg at a velocity of 4 m/second (assume the process is 100% efficient).

$$\begin{aligned} \text{Weight} &= 200 \times 9.81\text{N} \\ &= 1962\text{N} \\ W &= f \times d \\ &= 1962 \times 4 \\ &= 7848\text{Nm or } 7848\text{J} \\ P &= \frac{W}{t} \\ &= \frac{7848}{1} = 7848\text{J/s} = 7848 \text{ watts}\end{aligned}$$

Try this

An electric motor drives a pump which lifts 1 000 litres of water each minute to a tank 25m above the main storage tank. Ignoring efficiency, calculate the power rating of the motor.

_____

_____

_____

Figure 3.22 _Water pump_

Note

1 litre of water weighs 9.81N.

Practical problems involving efficiency

The efficiency of an electrical motor is the ratio of the electrical power output from the motor to its power input and is calculated by using:

$$\text{efficiency } (\eta) = \frac{\text{output power}}{\text{input power}} \times 100\%$$

Example:

Let's calculate the efficiency of a 230V motor that takes a current of 15A and has an output power of 2880 watts.

$$(\eta) = \frac{\text{output power}}{\text{input power}} \times 100\%$$

$$= \frac{2880}{230 \times 15} \times 100\%$$

$$= 83.5\%$$

Example:

A single-phase motor drives a pump which raises 600 litres of water per minute to the top of a building 10m high. Calculate the power that the motor must provide if the pump is 50% efficient.

$$\text{Weight} = 600 \times 9.81 \text{ N}$$

$$= 5886 \text{ N}$$

$$W = F \times d$$

$$= 5886 \text{ N} \times 10 \text{ m}$$

$$= 58860 \text{ Nm or } 58860 \text{ J}$$

$$\text{Pump output power} = \frac{58860}{60}$$

$$= 981 \text{ watts}$$

$$\text{Pump input power} = \frac{\text{pump output power}}{\eta} \times 100$$

$$= \frac{981}{50} \times 100$$

$$= 1962 \text{ watts}$$

$$\text{Motor output power} = \text{pump input power}$$

$$\therefore \text{Motor output power} = 1962 \text{ watts}$$

Try this

Calculate the output power rating of an electric hoist motor which is required to raise a load of 250kg at a velocity of 2 m/second. The efficiency of the hoist is 40%.

_____

_____

_____

Try this: Crossword

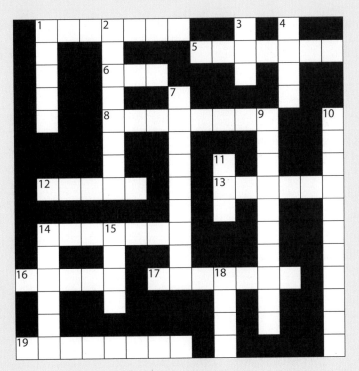

Across

1	Type of gear. (7)
5	A force of 9.81N is required to lift a load of 1kg against this? (7)
6 & 12	This shows statistical information. (3, 5)
8	Unit of resistivity. (3, 5)
13	Measured in newtons. (6)
14	Type of ammeter. (5, 2)
16	Actual power is measured in these. (5)
17	In a capacitive circuit the PF is? (7)
19	10^6 Watts. (8)

Down

1	Unit of inductance. (5)
2	Top heavy fraction. (8)
3	What will the current do with respect to the voltage in an inductive circuit? (3)
4	Trigonometric ratio. (4)
7	This electrical variable's symbol is Z. (9)
9	Ratio between output and input power. (10)
10	On the hypotenuse of a power triangle. (4, 7)
11	Power of ten for 100. (3)
14	What a capacitor stores in an electrical circuit. (6)
15	Gear teeth do this. (4)
18	Collected, sorted and analyzed. (4)

Congratulations you have now completed Chapter 3. Correctly complete the self-assessment questions before you progress to Chapter 4.

SELF ASSESSMENT

Circle the correct answers.

1   A bundle of conduit has a mass of 200kg. What force will be required to lift it?

   a.   200N

   b.   20.39N

   c.   1962.00N

   d.   1962kg

2   The effort required to lift a load of 40kg when using a four-pulley system is:

   a.   10kg

   b.   20kg

   c.   40kg

   d.   160kg

3   How much kinetic energy is possessed by a 2kg mass that is moving with a velocity of 4.5m/s?

   a.   9J

   b.   20.25J

   c.   40.5J

   d.   81J

4   What is the power output of a hoist that can lift a mass of 20kg a height of 15m in 40s?

   a.   7.5W

   b.   73.58W

   c.   7500W

   d.   12kW

5   A compound pulley system has:

   a.   one fixed pulley

   b.   one movable pulley

   c.   two fixed pulleys

   d.   one fixed and one movable pulley

Electrical science

4

RECAP

Before you start work on this chapter, complete the exercise below to ensure that you remember what you learned earlier.

1   Mass is the quantity of _____ that a body contains and is measured in _____.

2   What is the downward force on a bench of a motor with a mass of 150kg?

    _____

3   A force of 0.2N is used to move an object through 0.1m in 4.5s. Calculate the work done and the power.

    _____

4   Teeth on gears prevent _____ and they make it possible to determine _____
    gear _____.

LEARNING OBJECTIVES

On completion of this chapter you should be able to:

● Describe the principles of electron flow and current flow.

● Distinguish between materials which are good conductors and good insulators.

● State the types and properties of different electrical cables.

● Describe what is meant by resistance and resistivity in relation to electrical circuits.

● Calculate the values of electrical variables related to Ohm's law.

● Explain the relationship between current, voltage and resistance in series and parallel dc circuits.

● Calculate values of current, voltage and resistance in series and parallel dc circuits.

● Calculate values of power in series and parallel dc circuits.

● State what is meant by the term voltage drop in relation to electrical circuits.

● Describe the chemical and thermal effects of electrical currents.

Part 1 Electron theory

Have you ever considered why some materials are used to conduct electricity and others act as insulators? Some things, like water, can be used as either, depending on how pure they are. Everything we use is made up of molecules which contain atoms. It is the structure of these atoms that provides the answer.

Water is a good material to start with, because it contains two types of atoms, hydrogen and oxygen. The chemical formula for water is H_2O. This means it has twice as many hydrogen atoms as oxygen. Let's first consider hydrogen.

Hydrogen has the atomic number 1 and this means it has one set of components in its make up. All atoms consist of two main parts: the central nucleus, which contains the protons and neutrons, and the orbiting satellites, called electrons. The hydrogen atom has one proton and one electron. The electron revolves around the nucleus in a three-dimensional orbit.

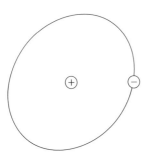

Figure 4.1 *Hydrogen atom*

As protons are positively charged, neutrons are neutral (neither positive nor negative) and electrons negatively charged, the atom overall has a neutral charge. All atoms in their neutral state have the same number of protons as electrons. The hydrogen atom is by far the simplest to follow with just one proton and one electron. Now let's look at an atom of a material that we know is a good electrical conductor, copper.

The copper atom, with an atomic number of 29, is far more complex than the hydrogen atom. Its nucleus consists of 29 protons (positive charges), and when it is in its neutral state the copper atom has 29 electrons. The electrons are spread out across four orbits, or shells, as shown in Figure 4.2.

In the outermost orbit there is only one electron, and this is shielded from the nucleus by the other electrons. The effect of this is to make the outer electron very loose in

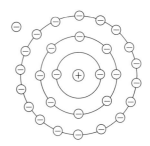

Figure 4.2 *Copper atom*

its shell and easily affected by other atoms. Remember that the electrons are on three-dimensional orbits and each orbit is an ellipse. This means that the outer electron is often closer to the nucleus of another atom than its own and so is attracted to the outer shell of that atom.

So, in copper, the outer electrons of the atoms move around from one atom to another. These are called free electrons. When an electron, which is negatively charged, leaves an atom, the 29 positively charged protons are stronger than the remaining electrons, so the atom becomes positively charged. This is referred to as a positive ion. In this state the atom is very attractive to negatively charged free electrons from other atoms. When copper is not being used as an electrical conductor the free electrons move about in a completely free fashion (randomly) in any direction throughout the metal.

Electron and current flow

The free negatively charged electrons, like all negative electrical charges, are attracted by positive charges.

Remember

Like charges repel and unlike charges attract.

An electric battery cell is basically a chemical unit which has two terminals. One of these has a surplus of electrons, the other a surplus of positively charged ions. When a copper conductor is connected across a cell the free electrons in the conductor are attracted to the positive terminal of the cell. This creates an electron flow in the conductor from negative to positive.

Conventional current flow →   ← Electron flow

Figure 4.3 *Electron and conventional current flow*

Although we now know that electron flow and current flow amount to the same thing, convention accepts that current flows from positive to negative.

This may be confusing, but remember that electrons are negative and are attracted to the positive. Current, we consider, is the opposite.

Remember

Never short out a battery or cell.

In practice we never short out a battery or cell with a conductor like copper. We need to have some kind of load connected, such as a lamp or a motor.

Try this

All atoms consist of two main component parts. The central nucleus contains the _____ and _____, and the orbiting satellites are called _____ .

_____ are charged positively, _____ are neutral and _____ are charged negatively.

Part 2 Conductors and insulators

We have seen that good electrical conductors, such as copper, have free electrons in their atoms. However, other materials will also conduct electricity, but not necessarily as well. Good conductors will pass current with the minimum of effort whereas other materials require greater pressure to be put on them. This pressure comes in the form of voltage, and it is this force that moves the electrons all together in one direction. This is why it is called the electromotive force (emf). If the pressure is applied at a high enough level, almost any material can be broken down into a conductor. Other examples of good conducting materials are aluminium, brass, carbon and silver.

Materials that will not conduct electricity at normal pressures are called electrical insulators. The atoms of insulating materials have their electrons tightly bound to them, therefore there will be no free electrons available to form an electric current, and none will flow.

Often the thickness of an insulator is related to the voltage it is designed to be used for. An example is the insulators for the cables that are carried on pylons across the country. Air is used to insulate the cables between pylons, and glass or ceramic insulators are used to keep the cables away from each other and the metal structure. Other examples of good insulating materials are: polyvinyl chloride (PVC), rubber, glass fibre and mica.

Try this

Good conductors will _____ current with the minimum of effort. An example of a good conductor material is _____ .

Materials that will not conduct electricity at normal pressures are called _____ .

All materials offer some _____ to current flow. Good _____ offer very little resistance under normal circumstances.

Types of electrical cables

There are many types of electrical cables available for various types of electrical installation wiring systems. We will have a look at some of the most commonly used cable types and the properties of their conductors, insulation and mechanical protection.

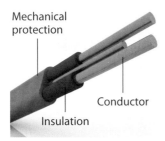

Figure 4.4 *The three main components of a cable*

Cable conductors

Copper or aluminium is widely used as the electrical conductor in many types of electrical cables. Aluminium is not as good a conductor as copper, having a resistivity about 1.6 times greater, but aluminium is much lighter than copper; the density of aluminium being less than a third of that for copper. Aluminium can be difficult to terminate as it oxidises easily and is prone to electrolytic corrosion.

Two mechanical properties that make copper and aluminium ideal materials for electrical conductors are that they are both ductile and malleable materials.

Ductile means that they can easily be extruded (drawn out into wires). Malleable means that they can be bent or shaped without cracking or breaking.

Annealed copper

A lot of cable conductors are made of annealed copper; this is copper that has been heat treated to make it tougher.

Remember

Low resistivity, or low resistance = good conductor.

Cable insulation

Plastic materials (polymers) are the most commonly used materials for cable insulation and they may be divided into two types: thermoplastic and thermosetting polymers.

Thermoplastic polymers soften when heated and solidify to their original state when cooled. There will be no damage to the polymer with repeated heating and cooling.

Thermosetting polymers become fluid when heated and change permanently to a solid state when cooled. The polymer may disintegrate with further heating.

Let's now look at some of the different cables and their properties, mainly focusing on the types of insulation and mechanical protection, having already covered the properties of copper and aluminium conductors which are the main two conducting materials used in many types of cable.

Types of cable and their properties

General purpose PVC-insulated PVC-sheathed cable

This type of cable is suitable for most types of domestic and commercial wiring installations where there is little risk of mechanical damage, extreme temperature or corrosion.

PVC sheathed and insulated – twin and multicore (Ref: 6242Y and 6243Y)

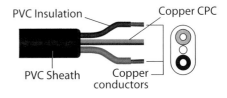

Figure 4.5 *Two-core & CPC (twin and earth) PVC/PVC cable*

PVC-insulated singles (Ref: 6491X)

This type of cable is suitable for commercial and industrial enclosed wiring installations. If used for domestic installations the cost would be considerably higher than using PVC/PVC cables.

Figure 4.6 *Single-core PVC-insulated copper cable*

PVC is a thermoplastic polymer and has the following properties:

- high electrical resistance
- impervious to water
- flexible and tough

- a maximum conductor operating temperature of 70°C (PVC insulation tends to soften above this temperature and at temperatures below 0°C it may become brittle and crack when handled)
- inexpensive compared to other types of cable
- singles have no mechanical protection; need enclosing in conduit or trunking (unless it is the type with a PVC sheath)
- the PVC sheathing gives light mechanical protection
- easy to terminate; no special terminations required
- plasticizers are added to make PVC more flexible; they can leach out and migrate into polystyrene foam materials and dissolve them; the cable itself will become degraded
- PVC exposed to sunlight, over a period of time, may be degraded by the UV radiation; black PVC sheathed cables are more resilient for outdoors.

Note

The reference numbers (e.g. 6242Y) are the British Approvals Service for Cables (BASEC) numbers which comply with specific standards for cable design.

Task

Referring to manufacturers' catalogues or websites, identify different types of flexible cords (cables) commonly used in domestic premises.

Part 3

PVC-insulated SWA-steel wire armoured PVC-sheathed cables (BS 6346)

These cables are widely used for all types of industrial installations where there is a greater risk of mechanical damage and corrosion. There are many different types that are very similar in structure; let's look at the XLPE SWA cable.

Cross-linked polyethylene (XLPE)

XLPE is a thermosetting compound which has better electrical properties than PVC.

Figure 4.7 *3-core XLPE insulated PVC bedded galvanized SWA PVC sheathed cable*

XLPE cables have grown in popularity for many types of installation because of their electrical, thermal and environmental properties. There are various types of XLPE cables manufactured; we will look at the properties of the armoured type:

- higher current-carrying capacity than PVC for the same size CSA cable
- suitable for medium voltage (MV) and high voltage (HV) systems as well as low voltage (LV) systems
- wider operating temperature range than PVC; maximum 90°C
- retains its mechanical strength at high temperatures long after PVC has gone soft
- galvanized steel wire armouring provides good mechanical protection, is corrosion resistant and may also be used as the circuit protective conductor (CPC) subject to it being of a suitable CSA
- the PVC outer sheath along with the armouring enables this cable to withstand more arduous environmental conditions, making it suitable for underground or overhead installations
- special armoured glands required for terminations.

Fire resistant cables

These cables are widely used for fire alarm and emergency lighting installations.

FP200 Silicon rubber insulated fire resistant cable

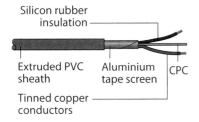

Figure 4.8 *FP200 Fire resistant cable*

Silicon rubber is a thermosetting compound that has the following properties:

- suitable where temperatures are very high; operating temperature 150°C
- even if the cable is completely burnt, the insulation forms a silica ash which is in itself an insulator, so

the cable may continue to function until the ash is disturbed, consequently, essential services such as fire alarm and emergency lighting systems can be maintained long after PVC insulation would have given up

- overall PVC sheath makes it suitable for damp and corrosive atmospheres
- aluminium tape acts as a metallic screen
- insulation is very soft so care must be taken not to damage it when terminating the cable.

FP400 is a fire resistant armoured version of this type of cable.

MIMS (mineral-insulated metal-sheathed) cable BS 6207

One of the most remarkable insulating materials is magnesium oxide (or magnesia). This is a white powder not unlike powdered chalk to look at.

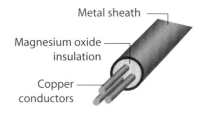

Figure 4.9 *Typical 3-core MIMS-mineral-insulated metal-sheathed cable*

MIMS cable is used in a wide range of installations. Although it is generally considered to be expensive for domestic wiring, you will find it used in factory workshops, churches, and in the dairy and food processing industries. It is frequently recommended for fire alarms and is sometimes the only type of cable suitable where environmental conditions are very severe. It has the following properties:

- can withstand extremely high temperatures
- virtually indestructible and keeps its insulating properties even though its temperature rises to well above 1 000 °C, after which the conductors will have melted and broken down

- mechanically very strong and robust; withstands knocks and bangs
- explosion proof as well as flameproof
- corrosion resistant with PVC oversheath
- seamless metal sheath of the cable is used as the circuit protective conductor (CPC), and if it is to be used with insulated fittings it must have an 'earth tail' incorporated in the termination
- although it is non-aging, it is expensive
- magnesium oxide must be kept completely dry, and since it can absorb moisture from the atmosphere, all terminations must incorporate a seal to act as a moisture barrier
- special gland, pot and seal assemblies are required for terminations
- both light duty and heavy duty types are available.

Task

Referring to manufacturers' catalogues or websites find the voltage ratings of the following cables:

Cable type	Voltage ratings
1 6242Y Flat PVC T&E	
2 6942XL SWA/XLPE/PVC	
3 FP200 gold	
4 MIMS (light duty)	
5 MIMS (heavy duty)	

Part 4

Aluminium tape (non-magnetic) armoured cable

Aluminium is widely used in armoured cables, PVC and XLPE, both as an armouring layer and as a conductor. Aluminium-armoured cable does not have the same current-carrying capacity as copper, but is very much lighter, and this can prove to be something of an advantage during installation.

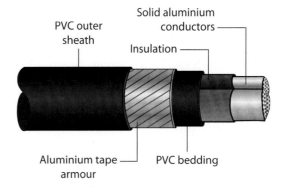

Figure 4.10 *Aluminium tape armoured cable*

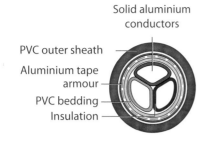

Solid aluminium conductors

PVC outer sheath

Aluminium tape armour

PVC bedding

Insulation

Figure 4.11 *Structure of aluminium tape-armoured cable*

Butyl rubber insulation

This is a synthetic rubber which is frequently used for high-temperature terminations such as heating elements. It will operate continuously without deterioration in temperatures up to 85°C.

For even higher temperatures (up to 185°C) glass fibre insulation may be used, and you can frequently find examples of this type of insulation on the internal wiring of heating appliances.

Voltage rating of cables

The voltage rating of a cable is the maximum voltage that may be continually applied to a cable (without any chance of the cable's insulation breaking down).

A cable may have a voltage rating 600/1 000V where U_0 = 600V, this is the nominal voltage between conductors and earth; U = 1 000V, this is the nominal voltage between line conductors.

Low toxicity

When looking at types of cable insulation we must consider the effects of smoke and fume emissions. When general purpose PVC cable burns, it can emit dense smoke and poisonous toxic gases (in particular hydrogen chloride gas) which can create serious health and safety hazards. These dangers are well known, and many contracts now carry a stipulation that low-toxicity, low smoke and fume (LSF) or low smoke halogen free (LSHF) cables have to be used.

When ordinary PVC cables burn, they can emit approximately 28% hydrogen chloride gas. Even though LSF type cable emissions are considerably lower, they can still give off smoke and toxic gases. LSHF type cables emit no

more than 0.5% hydrogen chloride gas and a lot less smoke than LSF cables.

There are now many different types of LSF and LSHF cables available; one example of a LSF cable is shown in Figure 4.12.

Figure 4.12 *4-core LSF XLPE SWA Cable*

Remember

For LSHF cables to be completely halogen free; the insulation and the sheath need to be halogen-free materials.

Fibre optic cables

Fibre optic cables are widely used for data cables. The conductor strands are made of optically pure glass as thin as a human hair (125µm or 0.125mm diameter), arranged in bundles (with many strands), and they are used to carry digital information (in the form of light signals) over very long distances.

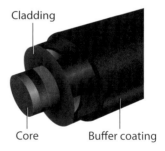

Cladding

Core   Buffer coating

Figure 4.13 *Typical fibre optic cable construction*

Fibre optic cable construction

1   Core – thin glass centre of the fibre where the light travels.

2   Cladding – outer optical material surrounding the core which reflects the light back into the core.

3   Outer covering (jacket) – plastic buffer coating that protects the fibres from damage and moisture.

Properties:

- less expensive than copper
- less signal loss than copper
- lower power rating

- non-flammable – no electric current through fibres
- lightweight and flexible.

These cables are commonly used for telecommunications and computer networks.

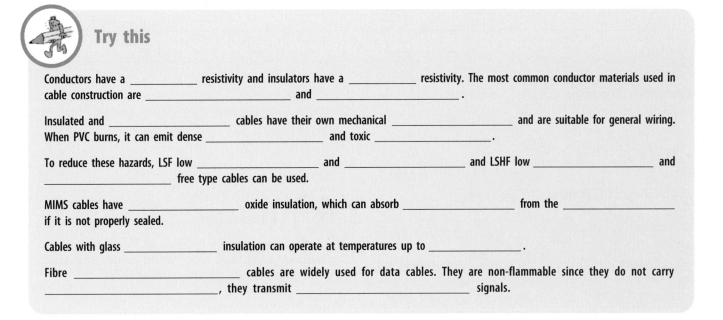

Try this

Conductors have a _____ resistivity and insulators have a _____ resistivity. The most common conductor materials used in cable construction are _____ and _____ .

Insulated and _____ cables have their own mechanical _____ and are suitable for general wiring. When PVC burns, it can emit dense _____ and toxic _____ .

To reduce these hazards, LSF low _____ and _____ and LSHF low _____ and _____ free type cables can be used.

MIMS cables have _____ oxide insulation, which can absorb _____ from the _____ if it is not properly sealed.

Cables with glass _____ insulation can operate at temperatures up to _____ .

Fibre _____ cables are widely used for data cables. They are non-flammable since they do not carry _____, they transmit _____ signals.

Part 5 Resistance and resistivity of electrical circuits

We have already seen that the resistance of electrical circuit conductors depends on the length, CSA, resistivity and the temperature of the conductors and that we can calculate the resistance of a conductor using the formula below (neglecting conductor temperature).

$$R = \frac{\rho\, l}{A}$$

Let's consider a simple circuit with a 20Ω resistor, for the load, connected to a 200V supply.

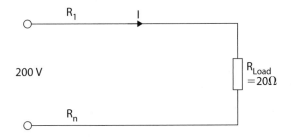

Figure 4.14 *Simple circuit*

The total resistance R_T of this circuit will also depend upon the resistance of the circuit conductors, therefore:
$R_{circuit} = R_1 + R_n + R_{Load}$

Example:

Calculate the total resistance of this circuit if the copper circuit conductors have a CSA of 1mm² and are each 5m long. The resistivity of the copper used is 17.2 μΩ mm.

Remember

All measurements should be of the same type, that is resistivity in micro ohm millimetres; length in millimetres; CSA in square millimetres. So we use 10^3 to convert metres to millimetres.

Try this

Determine the total resistance of the circuit in the example above if the circuit conductors are now each 20m long.

_____

_____

_____

Let conductor resistances:

$$R_1 + R_n = R_t \text{ and so } R_t = \frac{\rho l}{A} = \frac{17.2 \times 10^{-6} \times 10 \times 10^3}{1}$$
$$= 0.172 \, \Omega$$

The total circuit resistance:

$$R_{circuit} = R_t + R_{Load} = 0.172 + 20 = 20.172 \, \Omega$$

We can see that to find the total resistance of an electrical circuit, the resistance of the circuit conductors must also be included and this depends upon the resistivity, length and CSA of the conductors (also conductor resistance increases with increase in conductor temperature).

Relationship of voltage, current and resistance

The current that flows in a circuit is directly related to the pressure applied on the circuit by the voltage and the resistance that a circuit offers.

This relationship can be expressed by $\text{Current} = \frac{\text{Voltage}}{\text{Resistance}}$ or $I = \frac{V}{R}$

Remember

Ohm's law states: 'The current (I) in a circuit is proportional to the circuit voltage (V) and inversely proportional to the circuit resistance (R), at constant temperature.'

Example:

A circuit supplied with 200V and a load resistance of 10Ω will have a current flowing of $I = \frac{V}{I} = \frac{200}{10} = 10A$

There are occasions when the current is known and either the voltage or resistance needs to be calculated. In these cases the formula has to be re-arranged or transposed.

Try this

1 Determine the circuit current when a load of 25Ω resistance is supplied at 230V.

_____

_____

2 Determine the circuit current when 110V is applied to a load with a resistance of 22Ω.

_____

_____

Try this

Transpose the formula $V = I \times R$ to calculate the missing values in the grid below.

	Voltage	Current	Resistance
1.	240V	10A	
2.	12V	4A	
3.		22A	4Ω
4.	36V		25Ω
5.	240V	5A	

Series circuit

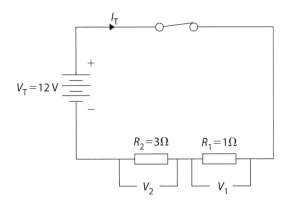

Figure 4.15

In the series circuit in Figure 4.15 the loads are shown using the general BS circuit symbol for resistors.

Given that the total voltage to the circuit is 12V and the resistors have values of 1Ω and 3Ω, the total circuit current and the potential differences (voltage drops) across the resistors can be calculated.

Example:

In a series circuit the total resistive path to the current flow is found by adding each of the resistors together. In this circuit the total resistance (R_T) is:

$$R_T = R_1 + R_2 = 1\Omega + 3\Omega = 4\Omega$$

We needed this value before the total circuit current could be calculated.

$$\text{Total current } I_T = \frac{V_T}{R_T} = \frac{12}{4} = 3A$$

Notice that to calculate total current the total voltage and total resistance must be used.

Now we can calculate the potential differences across R_1 and R_2. The current through each will be the same as the total current I_T, which is 3A.

Using $V = I\,R$ we first calculate for V_1

$V_1 = I_1 \times R_T$ and so $V_1 = 3 \times 1 = 3V$

As we only need the voltage across R_1 just the resistance used is R_1.

The voltage across R_2 can be calculated in two different ways, but the answer should be the same. First we can do it in a similar way to R_1, but using the value of R_2.

So $V_2 = I_2 \times R_2$ and so $V_2 = 3 \times 3 = 9V$

Or we could have taken the 3V calculated across R_1 and subtracted this from V_T.

In which case $V_2 = V_T - V_1$ and so $V_2 = 12 - 3 = 9V$

The answer is the same and it is acceptable to do it either way.

Remember

To find the total resistance of resistors connected in series, just add them together: $R_T = R_1 + R_2 + R_3$, etc.

Try this

In Figure 4.16 the total applied voltage is 10V and the circuit current is 4A. Calculate the value of each resistor when a potential difference of 2V is measured across R_1.

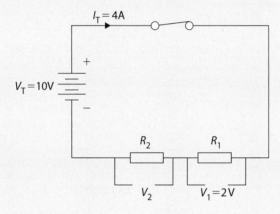

Figure 4.16 *Resistor values*

Part 6

Parallel circuits

Where there are only two resistors in parallel the total resistance may be calculated using the formula $R_T = \dfrac{R_1 \times R_2}{R_1 + R_2}$ which is referred to as the product ($\times$) over sum ($+$) method.

Where there are more than two resistors in parallel, use the formula:

$$\frac{1}{R_T} = \frac{1}{R_1} + \frac{1}{R_2} + \frac{1}{R_3} + \frac{1}{R_4} \dots \text{etc}$$

Example:

Determine the equivalent resistance of two 8Ω resistors connected in parallel.

$$R_T = \frac{R_1 \times R_2}{R_1 + R_2} = \frac{8 \times 8}{8 + 8} = \frac{64}{16} = 4\Omega$$

When two identical resistors are connected in parallel the equivalent resistance is halved.

Try this

For the circuit show in Figure 4.17 calculate:

1 The total resistance.

_____

_____

2 The total current.

_____

_____

3 The current in each resistance.

_____

_____

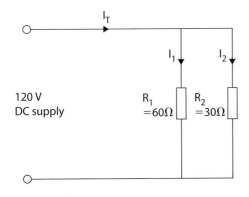

Figure 4.17 *Current and resistance*

Example:

Calculate (a) total resistance (b) total current and (c) the current in each resistance for the circuit shown in Fig 4.18.

a $\dfrac{1}{R_T} = \dfrac{1}{R_1} + \dfrac{1}{R_2} + \dfrac{1}{R_3} = \dfrac{1}{6} + \dfrac{1}{3} + \dfrac{1}{8} = \dfrac{4+8+3}{24}$

$\qquad = \dfrac{15}{24}$ so $R_T = \dfrac{24}{15} = 1.6\Omega$

b $I_T = \dfrac{V}{R_T} = \dfrac{24}{1.6} = 15A$

c $I_1 = \dfrac{V}{R_1} = \dfrac{24}{6} = 4\,A$ and $I_2 = \dfrac{V}{R_2}$

$\qquad = \dfrac{24}{3} = 8\,A$ and $I_3 = \dfrac{V}{R_3} = \dfrac{24}{8} = 3A$

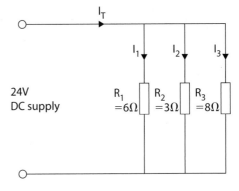

Figure 4.18 *Current and resistance (II)*

We can see that the sum of the currents in each parallel branch equals the total (supply) current: $I_T = I_1 + I_2 + I_3 = 4 + 8 + 3 = 15A$, and also note that the voltage across each resistor is the same as the supply voltage.

Combined series-parallel circuits

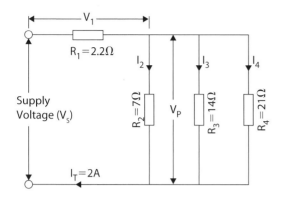

Figure 4.19 *Series-parallel circuit*

Example:

For the series-parallel circuit shown in Fig 4.19 calculate:

a   total resistance

b   supply voltage

c   current through each resistor.

a) First calculate the parallel resistance (R_p)

$$\frac{1}{R_p} = \frac{1}{R_2} + \frac{1}{R_3} + \frac{1}{R_4} = \frac{1}{7} + \frac{1}{14} + \frac{1}{21}$$

$$= \frac{6 + 3 + 2}{24} = \frac{11}{42} \text{ and so } R_p = \frac{42}{11} = 3.82\Omega$$

and the total resistance $R_T = R_1 + R_p = 2.2 + 3.82 = 6\Omega$

b) Supply voltage: $V_s = I \times R_T = 2 \times 6 = 12V$

c) $I_1 = I_T = 2A$

Voltage across R_1 $V_1 = I_T \times R_1 = 2 \times 2.2 = 4.4V$

Voltage across each parallel resistor:

$V_P = V_S - V_1 = 12 - 4.4 = 7.6V$

$I_2 = \dfrac{V_P}{R_2} = \dfrac{7.6}{7} = 1.1A$ and $I_3 = \dfrac{V_P}{R_3} = \dfrac{7.6}{14} = 0.54A$ and

$I_4 = \dfrac{V_P}{R_4} = \dfrac{7.6}{21} = 0.36A$

To confirm the result: $1.1 + 0.54 + 0.36 = 2A$ ✓

Try this

Three resistors of 4.8Ω, 8Ω and 12Ω are connected in parallel and are supplied at 48V dc. Calculate:

1   the current through each resistor

_____

_____

_____

2   the supply current

_____

_____

3   the total resistance of the circuit.

_____

_____

Part 7 Power in dc circuits

Earlier we saw that power was equal to the voltage times the current, $P = V \times I$. If we consider what we have seen with the relationship between voltage, current and resistance, there are two other ways of calculating power.

$P = I^2R$ when the current and resistance values are known

$P = \dfrac{V^2}{R}$ when the voltage and resistance values are known

Example:

Calculate the power dissipated by a 100Ω resistor when it is connected to a 24V dc supply.

$$I = \frac{V}{R} = \frac{24}{7} = 0.24A$$

$$P = VI = 24 \times 0.24 = 5.76W \text{ or}$$

$$P = I^2R = 0.24^2 \times 100 = 5.76W \text{ or}$$

$$P = \frac{V^2}{R} = \frac{24^2}{100} = 5.76W$$

Power in parallel and series dc circuits

Let's first consider a simple parallel circuit with two 12V lamps, each having a resistance of 2.4Ω, connected to a 12V dc supply, as shown in Figure 4.20.

Power dissipated by each lamp is: $P = \dfrac{V^2}{R} = \dfrac{12^2}{2.4} = 60W$

∴ total power dissipated is $P_T = P_A + P_B = 60 + 60 = 120W$

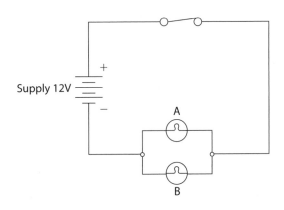

Figure 4.20 *Lamps in parallel*

Now let's connect these two 12V 60W lamps in **series** to the 12V dc supply as shown in Figure 4.21.

$$R_T = R_A + R_B = 2.4 + 2.4 = 4.8Ω$$

$$I = \frac{V}{R_T} = \frac{10^2}{4.8} = 2.5A$$

$$P_A = I^2R_A = 2.5^2 \times 2.4 = 15W \text{ and } P_B = P_A = 15W$$

∴ total power dissipated is $P_T = P_A + P_B = 15 + 15 = 30W$

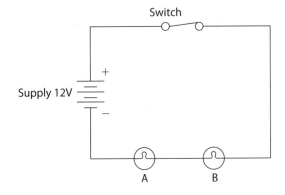

Figure 4.21 *Lamps in series*

Try this

Calculate the resistance of a 12V, 40W lamp, ignoring the temperature of the lamp.

We can see that the power dissipated by the same two lamps connected in series is one quarter of the power when the lamps were connected in parallel.

For any load to dissipate its rated output wattage, it must operate at its rated voltage.

Try this

The formula referred to as Ohm's Law is _____ .

In a series circuit the total resistive path to the current flow is found by _____ each of the resistors together.

In a series circuit the current will be the _____ through all the resistors and in a parallel circuit the _____ across each resistor is the same.

Power is equal to the voltage times the _____ .

Two other methods of obtaining the power in a series circuit are by using the value of the resistor in the formulae: _____

and _____ .

Part 8 Voltage drop

Voltage drop in an electrical circuit normally occurs when current flows through the circuit conductors. The greater the resistance of the circuit conductors, the higher the voltage drop, and consequently the voltage across the load will be less than the supply voltage. If the voltage drop is excessive the load may not work properly or safely and relays and contactors may not energize, fluorescents may not strike and motors may stall.

Consider the circuit in Figure 4.22 and we can see it is effectively a series circuit. The circuit conductor resistances R_1 and R_n and the lamp resistance R_L are all in series:

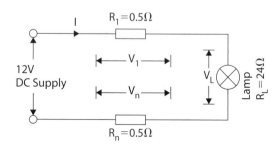

Figure 4.22 *Series circuit*

The total circuit resistance $R_T = R_1 + R_n + R_L = 0.5 + 0.5 + 24 = 25\Omega$

$$I = \frac{V}{R_T} = \frac{12}{25} = 0.48A$$

The voltage drop across conductor R_1 is: $V_1 = I \times R_1 = 0.48 \times 0.5 = 0.24V$ and the voltage drop across conductor R_n will be the same 0.24V.

Therefore the total voltage drop across the cable is $0.24 \times 2 = 0.48V$.

The voltage across the lamp is: $V_L = I \times R_L = 0.48 \times 24 = 11.52V$ and the sum of these three voltages = supply voltage: (0.24 + 0.24 + 11.52 = 12V).

Due to the cable resistance, we have lost 0.48V and the lamp has to work on 11.52V not on 12V. The 0.48V lost is the voltage drop due to the cable resistance.

Remember percentages: $\frac{11.52}{12} \times 100 = 96\%$

100 – 96 = 4%, hence, the voltage drop for this simple lighting circuit is 4%.

Note

To comply with BS 7671 the recommended maximum permissible voltage drop values for installations supplied from the public supply network are 3% for lighting circuits and 5% for power circuits. Hence, for a 230V single-phase supply, the voltage drop on a lighting circuit should not exceed 3% of 230V which is 6.9V.

Let's now check the power dissipated by the lamp.

$P = \dfrac{V^2}{R_L} = \dfrac{12^2}{24} = 6W$ will be the power without voltage drop.

With the voltage drop the power will be

$P = \dfrac{V^2}{R_L} = \dfrac{11.52^2}{24} = 5.3\,W$

6W – 5.3W = 0.7W loss of power, hence, the lamp will not work efficiently and will be less bright.

Try this

The circuit in Figure 4.22 has been modified and the cable is now six times longer. Determine:

1 the cable voltage drop

2 the voltage across the lamp.

Part 9 Effects of current flow

There are three main effects of current flow. These are:

- the production of heat
- the chemical effect
- the production of a magnetic field (we shall look at this in Chapter 5).

The heating (thermal) effect

Usually when we relate the amount of heat being produced by current flow we use the term 'power'.

An electrical appliance connected to a 230V supply and taking 5A consumes a power of 1 150W or 1.15kW.

The heating effect may be used as a room heater or to heat a filament up in a lamp so that light is given off. Heat is often a by-product of other effects, such as in an electric motor, where heat is a loss in the electromagnetic process.

The chemical effect

There are two main ways in which this can be seen.

The first is in batteries and cells, where the use of chemicals produces an emf which creates a current flow in a circuit.

Figure 4.23 *Primary cells*

In secondary cells external currents are used to replenish the chemicals used in the process so that they can be used again.

Figure 4.24 *A secondary cell*

The second way in which the chemical effect is used is in the electroplating industry. In this, metallic deposits can be plated on to surfaces by passing a current through electrodes and an electrolyte.

The electron flow through the electrolyte carries metal from the negative plate and deposits it on the positive plate. This process is known as electrolysis.

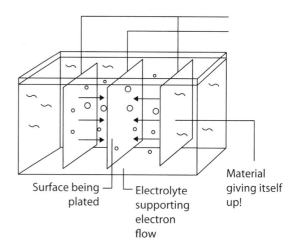

Surface being plated

Electrolyte supporting electron flow

Material giving itself up!

Figure 4.25 *Current passing through electrolyte*

Try this

List the three main effects of current flow.

_____

In the process called electrolysis the _____ flow through the _____ carries metal from the

_____ plate and deposits it on the _____ plate.

Congratulations you have now completed Chapter 4. Correctly complete the self-assessment questions before you progress to Chapter 5.

SELF ASSESSMENT

Circle the correct answers.

1   The number of electrons in a copper atom when it is in its neutral state is:

   a.   1

   b.   8

   c.   18

   d.   29.

2   Which of the following cable insulating materials has the highest temperature rating?

   a.   silicon rubber

   b.   fibreglass

   c.   Magnesium oxide

   d.   butyl rubber.

3   A circuit with a supply voltage of 110V has two 22Ω resistors connected in series. The circuit current will be:

   a.   0.4A

   b.   2.5A

   c.   5A

   d.   10A.

4   Two 12V 50W lamps are connected in series and fed from a 12V supply. The total power dissipated by the two lamps is:

   a.   25W

   b.   50W

   c.   100W

   d.   200W.

5   A resistive load is fed from a 24V dc supply. If the voltage drop across each circuit conductor is 0.6V the voltage across the load will be:

   a.   22.8V

   b.   23.4V

   c.   23.64V

   d.   24V.

5

Magnetism and electricity

RECAP

Before you start work on this chapter, complete the exercise below to ensure that you remember what you learned earlier.

1   Protons are _____ charged, electrons are _____ charged and neutrons are _____ .

2   LSHF type cables are low _____ halogen _____ .

3   In a series dc circuit the _____ remains the same and the _____ varies whilst in a parallel dc circuit the _____ remains the same and the _____ varies.

4   To comply with BS 7671, the recommended maximum voltage drop value for a 400V three-phase power circuit will be _____ .

LEARNING OBJECTIVES

On completion of this chapter you should be able to:

● Describe the magnetic effects of electric currents.

● Describe, with simple sketches, the pattern and direction of magnetic flux paths.

● State and apply the units, symbols and variables for magnetic flux and magnetic flux density.

● Determine the force on a current carrying conductor in a magnetic field.

● Determine the emf induced in a conductor moving through a magnetic field.

● Explain how an emf may be produced by self induction and mutual induction.

● Describe the basic principles of generating an ac supply.

● Explain the root mean square and average values of a sine wave.

Part 1 Magnetic fields

A magnet will affect the space around it in such a way that any other ferrous materials or magnets placed in this space experience force upon them. The actual space in which these forces occur is known as the magnetic field. The presence of a magnetic field surrounding a bar magnet can be demonstrated by sprinkling iron filings onto a sheet of paper (or thin card) placed on top of the magnet, and when it is gently tapped the iron filings take up the field pattern as shown in Figure 5.1.

Figure 5.1 *Magnetic field pattern of a permanent 'bar' magnet*

We can see that the iron filings produce a very random field pattern and that the flux does not exist as a number of separate lines. However, it is easier to explain and illustrate the various magnetic effects by representing magnetic field patterns as a number of separate lines of flux. This basic concept will be used throughout this chapter.

Consider the magnetic field pattern and flux paths of a permanent bar magnet and an electromagnet (solenoid) as shown in Figures 5.2 and 5.3.

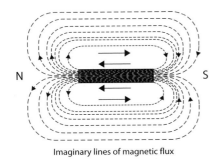

Imaginary lines of magnetic flux

Figure 5.2 *Typical bar magnet field pattern*

The flux paths that the lines of flux take are through the magnet and the space around the magnet. The lines of flux leave the north pole and enter at the south pole of the magnet.

The concentration (density) of the magnetic flux is strongest within the magnetic material since there are more lines of flux *closer together* than there are in the space surrounding the magnet.

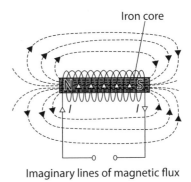

Imaginary lines of magnetic flux

Figure 5.3 *Typical electromagnet (solenoid) field pattern*

We can see that the field patterns and flux paths are very similar. The only difference being that the magnetic field of the solenoid is produced by the current flowing through the coil.

Looking at the diagrams more closely it can be seen that the imaginary lines of magnetic flux have very distinct properties:

● they always form complete closed loops

● they never cross one another

● they have a definite direction (north to south).

Two other properties which are not so distinct on the diagrams are:

● they try to contract as if they were stretched elastic threads

● they repel one another when lying side by side, and having the same direction.

Now let's consider the magnetic field patterns due to two bar magnets placed side by side. Figure 5.4 shows that when unlike poles are adjacent, the magnets are attracted to each other and Figure 5.5 shows that when like poles are adjacent the magnets are repelled from each other.

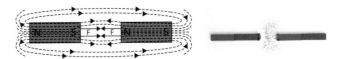

Figure 5.4 *Attraction*

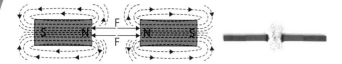

Figure 5.5 *Repulsion*

Concentric magnetic field directions (Maxwell's Corkscrew Rule)

When an electric current flows through a conductor it produces a magnetic field round the conductor in the form of concentric rings as shown in Figures 5.6 and 5.7.

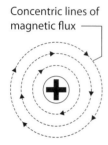

Concentric lines of magnetic flux

Figure 5.6 *Current flowing away from you (+) – concentric field direction is clockwise*

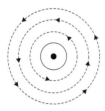

Figure 5.7 *Current flowing towards you (•) – concentric field direction is anticlockwise*

The current flow direction can easily be remembered by the right-hand grasp rule where the fingers point in the direction of the magnetic field and the thumb in the direction of the current flowing.

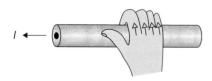

Figure 5.8 *Right-hand grasp rule*

Alternatively apply Maxwell's corkscrew rule, which states that, if a normal right-hand thread screw is driven along the conductor in the direction taken by the current, its direction of rotation will be the direction of the magnetic field.

Magnetic flux paths associated with adjacent parallel current carrying conductors

With current flowing in opposite directions as in Figure 5.9 the lines of flux tend to 'repel' each other as they are in the same direction. The conductors are forced away from each other (repulsion).

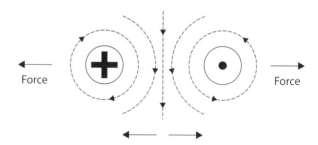

Figure 5.9 *Repulsion*

With current flowing in the same direction as in Figure 5.10 the lines of flux 'cancel' each other out. The conductors are now pulled towards each other (attraction).

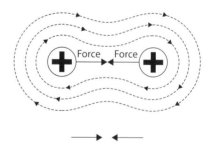

Figure 5.10 *Attraction*

Remember

A cross (+) indicates the current is flowing away from you and a dot (•) it is flowing towards you.

The right-hand grasp rule is used to find the relative direction of the current flowing and the magnetic field set up by the current.

Try this

Draw on the diagram below:

1   The resultant magnetic flux paths around the conductors and their directions.

2   The direction of the forces acting on each conductor.

Figure 5.11 *Flux paths and forces*

Right-hand grasp rule applied to a solenoid

The fingers point in the direction of the current flowing and the thumb indicates the direction of the magnetic field. Here the thumb points towards the north pole of the electromagnet. If the current direction is reversed then the field direction is also reversed.

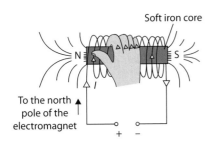

Figure 5.12 *Right-hand grasp rule*

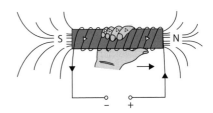

Figure 5.13

Try this

Draw the resultant magnetic field patterns on each diagram below.

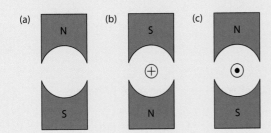

Figure 5.14 *Magnetic field patterns*

Try this

Unlike poles _____ and like poles _____ each other.

The fingers point in the direction of the _____ , and the thumb points in the direction of the _____ when applying the right-hand grasp rule to a straight conductor.

When applying the right-hand grasp rule to a solenoid the _____ point in the direction of the current flowing and the _____ indicates the direction of the _____ .

Part 2 Magnetic field properties

Magnetic flux (variable symbol Φ, unit symbol Wb)

Magnetic flux is a measure of the magnetic field produced by a permanent magnet or electromagnet. The unit of magnetic flux is the weber (Wb) which is pronounced 'vayber'.

Magnetic flux density (variable symbol *B* unit symbol T)

Magnetic flux density depends on the amount of magnetic flux (i.e. the number of lines of flux) which is concentrated in a given cross-sectional area of the flux path. The strength of a magnetic field is measured in terms of its flux density. The unit of magnetic flux density is the tesla (T) and 1 tesla = 1 weber per square metre.

This may be found by using the formula:

$$B = \frac{\Phi}{A} \text{ where } A \text{ is the area}$$

Therefore 1 tesla $= \dfrac{1 \text{ weber}}{1 \text{ square metre}}$

or 1 T $= 1 \text{ Wb/m}^2$

Figure 5.15 shows that magnetic flux density is a measure of the amount of flux (φ) which exists within an area (A) perpendicular to the direction of the flux.

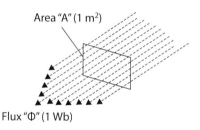

Area "A" (1 m²)

Flux "Φ" (1 Wb)

Figure 5.15 *Cross-section through a uniform magnetic flux*

Magnetic field strength

The strength of a magnetic field produced by, a solenoid for example, depends on:

● the magnitude of the current flowing through the solenoid's coil
● the number of turns on the coil
● the type of core material used.

Try this

Calculate the flux density existing in an area of 10m² if a uniform magnetic flux of 1Wb exists at right angles to that area.

_____

_____

The magnetic field strength can be increased by

- increasing the current flowing through the coil
- winding more turns on the coil

- selecting a suitable core material (with improved magnetic properties).

Try this

Magnetic flux (variable symbol _____) is a measure of the magnetic _____ and its unit is the _____.

Magnetic flux density (variable symbol _____) depends on the amount of magnetic flux and its unit is the _____ (T).

Magnetic field strength can be increased by _____ the current flowing through the coil, or by winding _____ turns on the coil.

Part 3

Electromagnetic force on a current-carrying conductor in a magnetic field

Nearly all motors work on the basic principle that when a current-carrying conductor is placed in a magnetic field it experiences a force. This electromagnetic force is shown in Figures 5.16, 5.17 and 5.18.

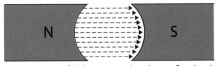

The magnetic field associated with two fixed poles

The magnetic field associated with a current-carrying conductor

The current is going away from you, therefore the field is clockwise (corkscrew rule).

Figure 5.16 *Electromagnetic force and a current-carrying conductor*

Now let's place the current-carrying conductor into the magnetic field. (Figure 5.17)

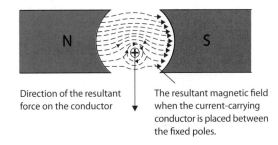

Direction of the resultant force on the conductor

The resultant magnetic field when the current-carrying conductor is placed between the fixed poles.

Figure 5.17 *Electromagnetic force on a current-carrying conductor in a magnetic field*

The main field now becomes distorted.

- The field is weaker below the conductor due to the fact that the two fields are in opposition.
- The field is stronger above the conductor because the two fields are in the same direction and aid each other. Consequently the force moves the conductor downwards.

If either the current through the conductor or the direction of the magnetic field between the poles is reversed, the force acting on the conductor tends to move it in the reverse direction. (Figure 5.18)

The direction in which a current-carrying conductor tends to move when it is placed in a magnetic field can be determined by Fleming's left-hand rule.

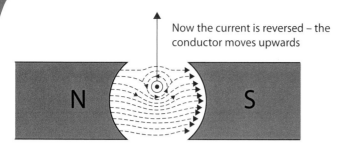

Now the current is reversed – the conductor moves upwards

Figure 5.18 *Reversing the field or current*

Fleming's left-hand (motor) rule

If the first finger, the second finger and the thumb of the left hand are held at right angles to each other (Figure 5.19) then with the first finger pointing in the direction of the field (N to S), and second finger pointing in the direction of the current in the conductor, then the thumb will indicate the direction in which the conductor tends to move.

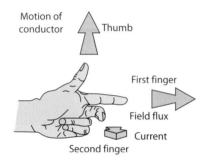

Figure 5.19 *Fleming's left-hand rule*

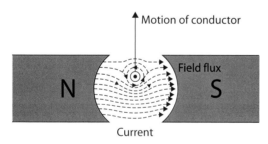

Figure 5.20 *Magnetic field and current flow*

Calculating the force on a current-carrying conductor in a magnetic field

The magnitude of the force on a current-carrying conductor can be calculated by the formula:

$$F = BIl$$

F = force in newtons (N)
B = flux density in tesla (T)
l = length of conductor affected by the magnetic field in metres (m)
and
I = current in amperes (A).

Note that the formula $F = BIl$ only applies to conductors which are moving at **right angles** to the magnetic field.

Example:

A conductor 0.3m long carries a current of 30A at right angles to a magnetic field of flux density 1.5T. Determine the force exerted on the conductor.

$$F = BIl$$
$$= 1.5 \times 0.3 \times 30$$
$$= 13.5N$$

Try this

Draw, on the diagram below, the field around each current-carrying conductor and the main field between the two poles. Also indicate the direction of the two forces acting on each conductor (by applying Fleming's left-hand rule) and the direction of rotation of the rotor.

Figure 5.21 *Simple motor with a single loop rotor coil*

Try this

A conductor 0.8m long is situated at right angles in a magnetic field of flux density 0.6T. Determine the current flowing in the conductor if it has a force of 150N exerted on it.

_____

_____

Part 4 Induced emf

A straight conductor passing between the poles of a magnet has an emf induced in it which is equivalent to the product of the flux density, the length of conductor in the field and its velocity.

$$E = Blv$$

E = induced emf in volts V

B = flux density in tesla T

l = length in metres m

v = velocity in metres/second m/s.

The direction of the induced emf can be determined by Fleming's right-hand rule.

Fleming's right-hand (generator) rule

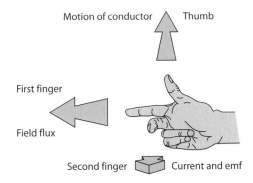

Figure 5.22 _Fleming's right-hand rule_

The formula $E = Blv$ can only be applied when the conductor is moving at right angles to the magnetic field, as shown in Figure 5.23.

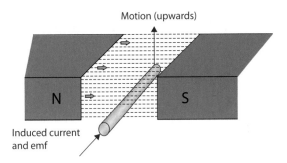

Figure 5.23 _Conductor moving at right angles to magnetic field_

Example:

When a conductor 0.2m long is moved at a velocity of 12m/s through a magnetic flux which has a strength of 0.9T it produces an emf of E = Blv, E = 0.9 × 0.2 × 12 E = 2.16V

Induced emf by rate of change of magnetic flux

$$E = \frac{N(\Phi_2 - \Phi_1)}{t}$$

E = induced e.m.f in volts V

N = number of coil turns

Φ_1 = initial flux in webers Wb

Φ_2 = final flux in webers Wb

t = time taken for change in flux (from Φ_1 to Φ_2) s

Try this

Calculate the emf induced in a conductor of 30cm length which is moving through a magnetic field of flux density 0.5T at a velocity of 40m/s.

Example:

The flux linking with a 600-turn coil changes from 60mWb to 120mWb in 100ms. Calculate the induced emf in the coil.

$$E = \frac{N(\Phi_2 - \Phi_1)}{t}$$
$$= \frac{600(120 - 60)}{100}$$
$$= \frac{600 \times 60}{100}$$
$$= 360V$$

Self inductance

As discussed earlier, a current-carrying conductor possesses its own magnetic field. When this conductor is wound into a coil, the field produced has the properties of a magnet. This becomes more pronounced if the coil is wound around an iron core. When the current is first switched on the magnetic field expands outwards, cutting the conductors and producing an induced emf. This has the effect of opposing the emf and producing the rise in flux density.

A German physicist, Heinrich Lenz said the same thing (in German) in about the year 1834 and this became known as Lenz's Law:

> The direction of an induced emf is always such that it tends to set up a current opposing the motion or the change in flux responsible for producing that emf.

An inductor is said to have an inductance of **one henry** when a current, which is changing at the rate of **one ampere per second**, produces an induced emf of **one volt**.

The henry is the unit of inductance and you will find it used to evaluate the inductive properties of chokes and coils in a variety of applications. One henry is quite a large value in practical terms and it is usual to find inductors measured in millihenries for everyday applications.

When an inductive circuit is first switched on, the induced emf opposes the increase in current. Instead of immediately rising to its final value, the current rises more gradually, at a rate governed by the inductance of the circuit.

Mutual inductance

The transformer is a static electromagnetic device which operates on the principle of mutual inductance. Having

Try this

A flux linking with a 300-turn coil changes from 2mWb to 4mWb in 0.02s. Calculate the magnitude of the induced emf.

studied electromagnetism you will be aware that a coil which is situated in its own magnetic field is capable of producing an induced emf due to its inductive properties.

If two coils share the same magnetic field, an emf will be induced in both coils. When a change in current in the first coil induces an emf in the second they are said to be mutually inductive.

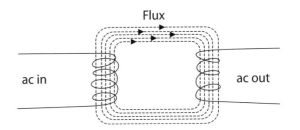

Figure 5.24 *Basic transformer*

Try this

Explain briefly by which method an emf will be produced in each of the coils, A and B, in Figure 5.25.

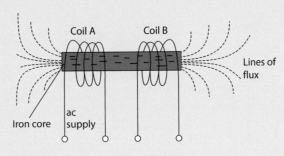

Figure 5.25 *Emf induction*

_____

_____

Try this

To calculate the induced emf in a conductor moving at right-angles to a magnetic field, apply the formula _____ .

An emf may be produced by _____ induction or _____ induction.

An inductor has an inductance of 1 _____ when a current which is changing at the rate of 1 _____ produces an induced emf of _____ .

Part 5 Generating an alternating current (ac) supply

We established earlier that when a conductor moves through a magnetic field a current and an emf is induced in the conductor.

Let's now form this conductor into a single loop coil and fit it to a shaft on bearings so that it can rotate between two magnetic poles as shown in Figure 5.26.

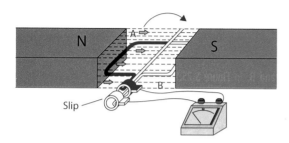

Figure 5.26 *Simple single-loop ac generator*

To do this the connections to the coil must be made through rings (slip rings) that are fitted to the shaft and slide against fixed contacts (carbon brushes) connected to the meter. As the coil rotates, each of the two long coil sides, A and B, goes past the poles in turn. At the instant the coil is as shown, both coil sides A and B are cutting through the lines of magnetic flux at the maximum rate, since they are both moving at right angles to the lines of magnetic flux. This produces a maximum emf and maximum deflection on the meter to the left.

As the coil rotates, the emf drops off until the coil is vertical between the poles. At this point the emf is zero, since the coil sides are moving parallel to the lines of magnetic flux, therefore no flux is being cut.

The coil now continues to rotate and the emf builds up again, but in the opposite direction. The effect of rotating the coil is to continually change the direction of the current flowing in the circuit. This is why this system of generation is known as alternating current (ac). As there is only one output from this type of generator it is known as a single-phase supply.

The speed at which the coil rotates is very important as this governs the frequency, or cycles per second, of the supply. In the UK and Europe the frequency is 50 hertz (50 Hz). This means that there are 50 cycles each second. The coil in the generator we have looked at has to turn 50 complete revolutions each second to achieve this.

The sequence that the current goes through in one cycle, one 360° turn of the coil, is called a waveform. The type of wave this produces is a **sinusoidal or sine wave** as shown in Figure 5.27. This waveform is the same as that of the voltage which appears during the generation of current within the coil.

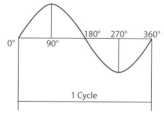 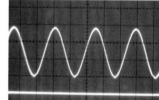

Figure 5.27 *Typical sine wave*

![Try this icon] **Try this**

1   What are the slip rings used for? _____

2   The number of degrees a coil needs to rotate to create one complete sine wave is _____.

3   Referring to Figure 5.26, apply Fleming's right-hand (generator) rule to determine the current direction in each coil side A and B.

_____

_____

Three-phase generators

Although single-phase is used for domestic supplies, it is seldom generated in large quantities. Power stations use generators (or, as we should call them, alternators) that produce a three-phase output.

The principle of a three-phase alternator is similar to that of a single-phase device, but the construction is quite different. This time the generating coils remain stationary while the magnetic field revolves. In practice the moving magnet may be an electromagnet fed from a dc supply, but we shall show it as a permanent magnet to keep it simple.

As the magnet shown in Figure 5.28 rotates past each coil in turn, an emf is generated in that coil. As the coils are 120° apart, the sine wave generated in each is 120° 'out of phase' with the next one. This creates three waveforms starting 120° after each other.

A three-phase supply can be used in many different ways and can be very adaptable. It is particularly suitable for large power applications.

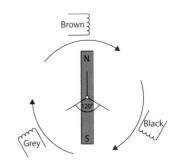

Figure 5.28 *Three-phase alternator*

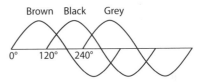

Figure 5.29 *Sine wave phases*

Sinusoidal waveform (sine wave) characteristics

RMS and peak or maximum value

The public ac supply of 230V is not the maximum voltage that is applied. It is a value of voltage that will give an equivalent amount of power to a 230V dc supply.

The maximum or peak voltage is about 325.22V.

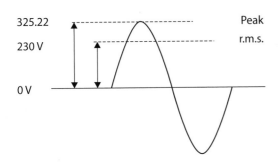

Figure 5.30 *Peak and RMS values*

The RMS (root mean square) value, known as the working voltage, is 0.707 of the peak (maximum) value. Or in other words the peak value is 1.414 times the RMS value.

The RMS values of both voltage and current are the most effective and useful values of the ac waveform. If we connect a resistive load to a steady dc supply and draw a dc current equal to the ac RMS current value, the heating effect will be the same in both cases or the useful power dissipated will be the same in both cases. For example:

13A (RMS) = 13A (dc) or 60W (RMS) = 60W (dc)

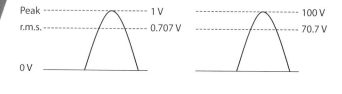

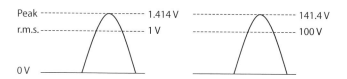

Figure 5.31 *Relationship between peak and RMS values shown over a half cycle for clarity*

Average values

Over one complete cycle of 360° the average value is zero. This is calculated by taking the area of the positive half-cycle and taking away the area of the negative half. As these are the same but in opposite directions they cancel each other out.

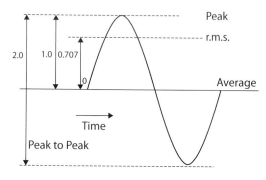

Figure 5.32 *Positive and negative half cycles*

We sometimes need to know the average value of one half cycle. As the shape is not symmetrical the average is not 0.5 of the peak – it is in fact **0.637**.

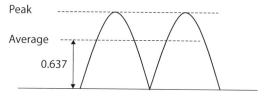

Figure 5.33 *Average over half a cycle*

Example 1:

A sinusoidal current has a maximum value of 28.3A. Calculate (a) its RMS and (b) average value.

a $I_{RMS} = I_{max} \times 0.707 = 28.3 \times 0.707 = 20$ A
b $I_{av} = I_{max} \times 0.637 = 28.3 \times 0.637 = 18$ A

Example 2:

Calculate the peak and average voltages of a sinusoidal supply with an RMS value of 50V.

$$V_{rms} = V_{max} \times 0.707 \text{ and so } V_{max} = \frac{V_{rms}}{0.707} = \frac{50}{0.707}$$
$$= 70.72V$$

$$V_{av} = V_{max} \times 0.637 = 70.72 \times 0.637 = 45V$$

Try this

Calculate a) the maximum and b) average values of a sinusoidal current which has an RMS value 14A.

Congratulations you have now completed Chapter 5. Correctly complete the self-assessment questions before you progress to Chapter 6.

SELF ASSESSMENT

Circle the correct answers.

1   The unit of magnetic flux density is the:

   a.   weber
   b.   henry
   c.   tesla
   d.   ampere-turn

2   The output from a rotating coil of a single-phase alternator is by:

   a.   permanent magnet
   b.   slip rings
   c.   springs and conductors
   d.   continuous connection by wire.

3   When applying Fleming's left-hand rule, the thumb indicates:

   a.   field direction
   b.   induced current direction
   c.   motion direction
   d.   induced emf direction.

4   The RMS voltage is found to be 200V. This means that the peak voltage must be approximately:

   a.   282V
   b.   200V
   c.   141V
   d.   127V.

5   The maximum value of a sinusoidal current is 25A; its average value will be approximately:

   a.   16A
   b.   17.5A
   c.   35A
   d.   39A.

6

Supply and distribution systems

RECAP

Before you start work on this chapter, complete the exercise below to ensure that you remember what you learned earlier.

1 Applying Fleming's right-hand rule, first finger is _____ direction, second finger is _____ and _____ direction, thumb is direction of _____.

2 We have seen that to produce an emf there must be relative _____ between a conductor and a _____ field.

The simple single-loop generator works on the principle of moving a _____ between fixed 'magnetic field' _____ to induce an emf.

The transformer works on the principle of a changing _____ field inducing an emf into the stationary transformer _____.

3 For an alternating current 0.707 is the _____ value and 0.637 is the _____ value.

4 Briefly explain why the RMS value of a sinusoidal waveform is the most effective value of the waveform.

_____

LEARNING OBJECTIVES

On completion of this chapter you should be able to:

● Describe how electricity is generated, transmitted and distributed.

● Specify the features and characteristics of generation, transmission and distribution systems.

● Explain other sources of generation.

● Describe the main characteristics of single and three-phase supplies.

● Describe the operating principles, applications and limitations of transformers.

● State the types of transformers used for supply and distribution.

● Determine by calculation and measurement primary and secondary voltages and current and kVA rating of transformers.

Part 1 Electricity generation

Power stations

Electricity is generated by rotating the shaft of an alternator (ac generator). The energy required to drive an alternator can be produced by steam plants and natural resources.

Steam plants

Most electricity is generated by producing heat through an energy conversion process. The heat is then used to change water into steam and the steam drives turbines which in turn rotate the alternator.

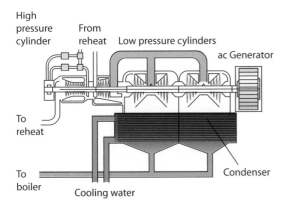

Figure 6.1 *Steam turbine generator plant*

Fossil-fuel power station

A fossil-fuel power station burns fossil fuels such as coal, oil or natural gas to produce electricity. Natural gas is the cleanest fossil fuel, emitting less harmful pollutants into the atmosphere.

Coal has been used more than any other fuel to produce the heat in power stations. A station with an output of 2 000MW consumes about 5 million tonnes of coal a year. For this reason coal-fired power stations are often sited near to coal mines or on easy transport routes, such as road, rail or sea.

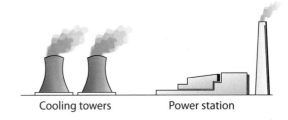

Figure 6.2 *Coal-fired power station.*

It is necessary to have high temperatures to create the steam, but it is also necessary to cool parts of the system down. For this reason wherever possible power stations are situated close to a large source of water, such as the sea or a large river. A coal-fired power station will use billions of litres of cooling water each year.

Oil can be used as a fuel for steam-raising plants. Fuel oil is a by-product of the refining process and for reasons of economy, oil-fired power stations may be situated close to refineries or adjacent to tanker terminals.

Nuclear power stations

Nuclear power stations also use the steam process. The siting of these stations relates more to possible hazards from the nuclear fuel than from its transportation. In comparison with coal the amount of fuel used is very small – about 4.5 tonnes of uranium each week for a 1 000MW station.

The cooling, however, still has the same requirements as coal stations, and the disposal of the waste radioactive materials is an additional problem.

Combined cycle gas turbine (CCGT) power stations

CCGT power stations have a gas turbine generator to generate electricity and heat from its exhaust is used to make steam, which in turn drives a steam turbine to generate additional electricity, thus giving a better overall efficiency from the power plant.

Natural resources

The most widely used natural resource for generating electricity is **hydro-electric power**. Although electricity is cheap to produce in this way it often has to be transmitted over long distances to the nearest populated and industrial areas. To produce electricity using hydro-electric power a head of water has to be available. Figure 6.4 shows how reservoirs high in the mountains are used to create pressure through an artificial tunnel. By controlling the flow in the tunnel the output can be varied throughout the day.

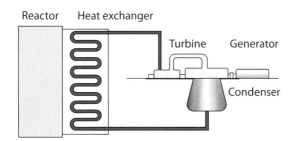

Figure 6.3 *Nuclear power station*

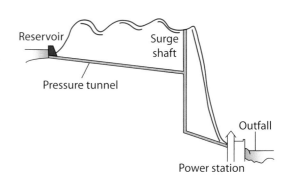

Figure 6.4 *Typical hydro-electric station*

Gas turbine power stations

Gas turbines, instead of heating steam to turn a turbine, use hot gases from burning natural gas to turn the turbine and generate electricity.

Try this

Electricity is generated by rotating the shaft of an _____ .

_____ fuel power stations burn _____ , natural gas or _____ to produce electricity.

Hydro-electricity is _____ to produce.

Part 2 Electricity generation and the grid system

Generation to transmission

The output voltage of a generator set in a modern power station is 25kV ac, and this voltage is then transformed to 400kV (or 275kV in some situations) for transmission purposes.

Transmission switching stations

When the electricity comes from the generator it passes through a switching station which transforms it up to the transmission voltages.

The switching station, as the name implies, is more than just a transformer. As with all circuits there must be means of overcurrent protection and means of isolation. A circuit breaker provides the means of overcurrent protection, and is designed to operate in a fraction of a second under overcurrent conditions. The isolators are not designed for switching when the load is connected, and can only be used when the circuit breaker has already disconnected the supply. As can be seen from Figure 6.5 there are two isolators. These can be used so that the circuit breaker can be completely isolated from the grid or the transformer, both of which may be 'live'. This enables the circuit breaker to be maintained without any danger from the supply.

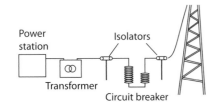

Figure 6.5 *Typical switching station*

Transmission

The majority of electrical power is transmitted on overhead cables suspended from pylons. Although the pylons can look unsightly it is by far the most cost-effective way of transmitting power. High voltages are used for transmission and these have to be well insulated from one another and from earth.

When bare conductors are placed high in the air, as in Figure 6.6, and spaced apart, then the space between the conductors and the conductors and earth becomes the insulation and no further covering is required. The air around the conductors also helps to keep them cool when they are carrying heavy loads.

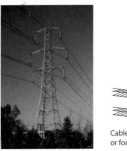

Cables are bunched in twos or fours as shown

Figure 6.6 *Overhead transmission*

To install high-voltage transmission cables underground is very costly. The cables need special cooling methods, such as oil being pumped through them, as the natural heat dissipation into the surrounding soil is not sufficient. They also require very thick insulation to protect the conductors from each other and the earth around them. If a fault develops on the cables when they are exposed in the air, this can be seen and often it is a comparatively straightforward job to put it right. However, if a fault develops on an underground cable this is not readily accessible and can take time and cost far more to repair. Having said this, 400kV cables are placed underground when there are no alternatives or it is important for environmental reasons, and in the UK there are more of these cables underground than anywhere else in the world.

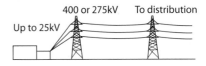

Figure 6.7 *Transmission voltages*

The transmission system

When the 132kV transmission network was first established it was known as the standard grid system. As demand grew the 275kV super-grid system was constructed and we now have an integrated 400/275kV main transmission super-grid system. This consists of a latticework of cables which connect power stations and large load areas together.

In addition to the transmission system in the UK there are cross-channel links with the French supply system. These are underwater cables going from the Kent coast to north east France. The time and lifestyle differences in the two countries are factors which make these links practicable. The peak demand times vary, and this allows the UK to import power from France when demand requires. It also allows the export of energy to take place when the French demand is high. For practical reasons these cross-channel cables are supplied with dc and this is converted to ac at each end.

Remember

Electricity is generated as alternating current.

This means that it can be transformed up or down for transmission and distributed to where it is required.

Try this

A power station generator's output voltage is _____ .

Standard grid voltage is _____ , super-grid voltages are _____ and _____ .

Part 3 Distribution

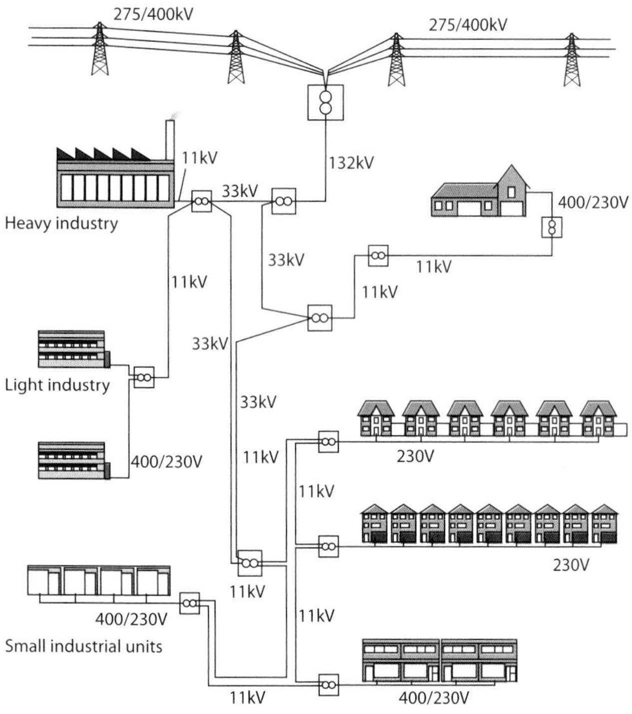

The distribution system can be split into three main sections:

● industrial
● commercial and domestic
● rural.

The reasons for splitting these are the different voltages they require and the remoteness of rural supplies. Wherever possible, distribution cables are laid underground. Apart from rural areas, almost all 11kV and 400/230V cables are buried underground, and a significant number of the higher voltage cables are now also buried.

The electricity suppliers have a legal responsibility to keep the supply within certain limits. Following voltage standardization in Europe these are, for voltage, a nominal supply of 400/230V, +10%, –6%; and for frequency it must not be more or less than 1% of 50Hz during a 24-hour period.

HV primary substation

An HV primary substation fed at 132kV or 33kV supplies 11kV distribution substations on a ring main system to supply small industrial, commercial and domestic consumers. A radial distribution system may be used; however,

Figure 6.8 *Distribution system showing the voltages used*

Try this

A public electricity supply company has to comply with the legal requirements and their stated nominal voltage is 230V. Calculate the supply's

1 highest voltage _____

2 lowest voltage _____

the ring distribution system offers greater security of the supply to consumers with each substation being fed from two directions. Automatic circuit breakers provide graded overcurrent protection on the incoming and outgoing sides of each transformer.

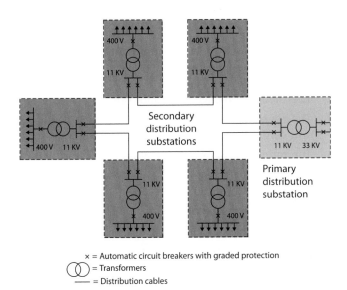

Figure 6.9 *HV ring main distribution system*

An HV/LV, 11kV to 400/230V, distribution substation supplies power from the transmission system to the distribution system of a local area.

HV/LV distribution substation transformer windings

The 11kV input to the transformer is connected in 'delta' whereas the 400/230V output is a 'star' arrangement. To

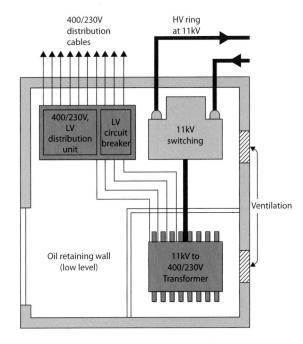

Figure 6.10 *Typical HV/LV substation layout.*

supply the delta connected windings a three-phase three-wire system is used, with no neutral conductor. The star connected output uses a three-phase four-wire connection with the centre point of the star being neutral and connected to earth.

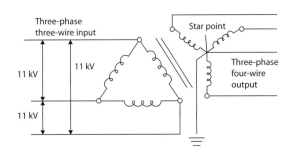

Figure 6.11 *Substation transformer windings*

Connected in delta the voltage across each of the lines is the same as the transformer winding, whereas a transformer winding connected in star will give us a voltage between phases of 400V and 230V between any phase and the neutral star point.

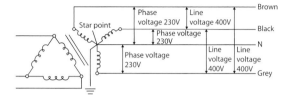

Figure 6.12 *Voltages available from a star connected transformer winding*

Figure 6.13 shows the different types of supply which can be obtained from the secondary side of a delta/star transformer.

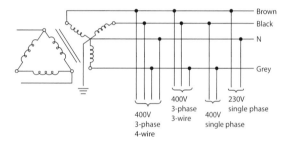

Figure 6.13 *Typical distribution arrangements*

Typical uses for each type of supply include:

400V 3-phase 4-wire – industrial, agricultural and commercial premises

400V 3-phase 3-wire – motor circuits

400V single-phase – welding plant

230V single-phase – small commercial and domestic premises.

Earth-fault loop path

In the event of a fault occurring between a live conductor and exposed metalwork, enough current has to flow through the conductor to make the protection device operate rapidly, therefore the total loop impedance, Z_S must be a low value. Figure 6.14 shows the path that a fault current, I, would take. The effect is to short the transformer out so that high currents are drawn from the system. If the protection device does not operate very fast the energy used may cause the cables to overheat, melt and start a fire, and there is a risk of serious electric shock while the exposed metalwork is live.

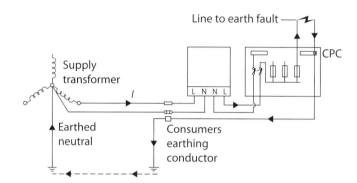

Figure 6.14 *Simple diagram of a TT System earth fault loop path*

Referring to Figure 6.14 and starting at the fault, the loop path comprises:

1   The circuit protective conductor (CPC).
2   The consumer's earthing terminal, earthing conductor and earth electrode.
3   The return path through the mass of earth for this TT system.
4   Supply earth electrode.
5   The earthed neutral of the supply transformer.
6   The transformer winding.
7   The line conductor from the transformer to the fault.

Remember

For rapid disconnection, the total loop impedance must be low to allow a high fault current to flow round the loop. If not, there is a risk of electric shock from the live exposed metalwork.

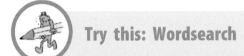

Try this: Wordsearch

Find the words from the list below in the wordsearch puzzle:

A	H	Y	D	R	O	E	L	E	T	R	I	C	H	S
S	E	O	D	I	S	T	R	I	B	U	T	I	O	N
U	E	G	N	F	G	I	N	E	N	F	S	T	S	P
B	N	L	S	B	S	N	U	C	L	E	A	R	U	O
S	P	O	E	T	E	V	O	S	N	T	B	B	P	W
T	G	H	I	U	O	S	I	A	L	W	A	C	E	E
A	T	I	S	T	F	S	H	E	P	L	N	G	R	R
T	D	T	P	I	A	L	D	O	I	N	S	I	G	S
I	T	I	N	H	S	R	I	S	H	S	B	C	R	T
O	F	O	U	G	I	H	E	S	I	O	N	D	I	A
N	U	N	D	D	S	A	D	N	S	T	F	D	D	T
F	L	D	P	L	H	S	N	T	E	O	S	I	P	I
I	N	D	U	S	T	R	I	A	L	G	F	E	S	O
N	T	S	I	V	S	T	A	R	P	O	I	N	T	N
T	R	A	N	S	M	I	S	S	I	O	N	P	I	H

Delta	Generation	Nuclear	Star Point	Supergrid
Distribution	Hydro Electric	Power Station	Substation	Transmission
Fossil Fuel	Industrial			

Part 4 Electrical generation from other sources

Solar photovoltaic systems

Photovoltaic cells, generally referred to as PV cells, convert light into electrical energy. However, the output from a single PV cell is quite small so they are connected in arrays to create a 'solar panel' with a higher output.

Solar thermal heating

A solar thermal heating system uses the heat generated from sunlight to heat water. The system uses an indirect cylinder to transfer the heated water within the solar collector to the consumer's hot water.

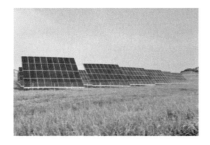

Figure 6.15 *Typical ground-located PV system*

Wind generation (micro and macro)

Wind generation is used for small-scale generation (micro) and large-scale generation (macro). Micro generation systems have small generation units and wind generators which are used for individual dwellings, industrial and commercial buildings. Macro generation relates to the larger wind turbines used for wind farms supplying electrical energy into the transmission and distribution system. Midi or small turbines are generally those in the region of 2–8kW.

Figure 6.16 *Macro wind generation*

Wave energy

Wave energy or wave power is a clean renewable form of electricity generation.

Ocean surface waves are caused by the wind as it blows across the sea. Wave energy is transported by these ocean surface waves. There are several methods used to get energy from waves, one simple method is shown in Figure 6.17 below.

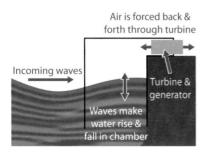

Figure 6.17 *Simple wave generator system*

The incoming waves cause water in a chamber to rise and fall; this causes air to be forced in and out of a hole in the top of the chamber. A turbine in this hole is turned by the air rushing in and out and drives a generator.

Micro-hydro generation

The term micro-hydro is used for hydro-electric generating systems which produce up to about 100kW output. These can be used to supply dwellings or small industrial/commercial installations, particularly those in remote areas.

Ground source heat pumps (GSHP)

GSHPs take low grade heat from the ground. At a depth below one metre there is a stable source of heat throughout the year. This low grade heat is then passed through an evaporator to the refrigerant in the heat pump. The refrigerant is then compressed, by an electrically driven compressor, which greatly increases its temperature. This high grade heat is then passed to the building to provide heating and hot water.

Combined heat and power (CHP)

In simple terms combined heat and power (CHP) uses the heat produced during the generation of electricity. The traditional electrical generation system produces heat and this is generally vented to the atmosphere, the steam clouds above power station cooling towers are an example of this. CHP uses the heat produced and rather than wasting it uses it to provide heating and hot water for consumers.

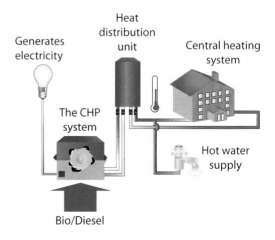

Figure 6.18 *Typical CHP system*

CHP units may be used to provide supplies for individual buildings such as a school and such a unit may reasonably deliver an electrical supply of 12–15kW and a heating supply of 26kW. This may be powered using the existing natural gas supply replacing the traditional gas boiler.

Micro CHP

The micro CHP units used for individual dwellings are generally smaller and provide an output in the region of 1kW for use in the home. The units normally produce electricity when heat is required and any electrical energy not used in the property may be sold back to the local network.

Batteries and cells

Most of the direct current (dc) we use is from batteries and cells. A cell stores electrical energy in a chemical form. Chemical action inside the cell produces an emf (voltage) and after a limited time the chemical action deteriorates and the voltage drops below a useful level. Eventually the chemical action ceases and the cell is of no further use (unless it is a rechargeable type).

Cell construction

A simple cell has two plates made of different materials, a positive plate and a negative plate immersed in an electrolyte. When the two plates are immersed in the

electrolyte it causes a chemical reaction to take place. A simple cell of this type can be made using a zinc plate and a copper plate immersed in a liquid electrolyte containing sulphuric acid and water.

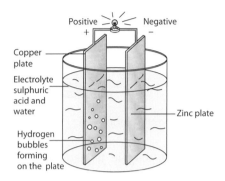

Figure 6.19 *A simple or voltaic cell*

Cells are made from different materials to give various voltages. Some examples are shown in Table 6.1.

Cells connected in series

Cells are often connected in series to give a higher voltage output. For example, a 9V battery contains six 1.5V cells connected in series in a single casing.

Table 6.1 *Cell properties*

Cell type	Positive material	Negative material	Electrolytes	Voltage produced
Voltaic	Copper	Zinc	Dilute sulphuric acid	0.5–1.0V
Law	Carbon	Zinc	Ammonium chloride	0.5–1.0V
Leclanché	Carbon	Zinc	Ammonium chloride	1.4V
Danielle	Copper	Zinc	Copper sulphate & zinc sulphate	1.17V
Weston Standard	Mercury	Cadmium	Cadmium sulphate	1.083V
Zinc carbon	Carbon	Zinc	Ammonium chloride & zinc	1.5V
Alkaline	Zinc powder	Manganese	Potassium hydroxide	1.5V
Mercury button	Zinc powder	Mercuric oxide	Alkaline	1.5V

Figure 6.20 *Single cell symbol*

The positive pole of one cell is connected to the negative pole of the other cell when two cells are connected in series (Figure 6.21).

Figure 6.21 *Series cell symbol*

Cells connected like this are termed batteries, although this name is also often used for single cells. A car battery has a set of cells linked together in series (inside an acid-resistant container) to produce 12volts. Each cell can produce approximately 2volts; therefore the car battery has six cells.

Figure 6.22 *Battery symbol*

Cells connected in parallel

Cells can also be connected in parallel; the two positive poles are connected together and the two negative poles are connected together with the output voltage remaining the same.

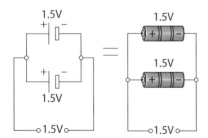

Figure 6.23 *Cells in parallel*

Cells connected in parallel can provide the same power for twice as long since they share the load. In effect this makes a 'larger' battery of the same voltage. This arrangement is used where a long life is important.

Where high voltage and high current are necessary, cells are connected in a series-parallel arrangement.

Remember

When cells are connected in series the voltages are added together.

When cells are connected in parallel the voltage is the same as a single cell.

Primary and secondary cells

Primary and secondary cells look similar; however, they are in fact very different. When a primary cell is connected to a circuit it produces electricity from a chemical reaction taking place within the cell; a secondary cell uses chemical action, but actually stores electricity in a chemical form. When the chemical action of a primary cell slows down and ceases the cell is dead and must be replaced. The secondary cell can be recharged with electricity many times, giving it a very long life. Recharging actually reverses the chemical process in the battery.

The lead acid cell

This is the most common type of secondary cell. The negative plate is formed from a spongy lead and the positive plate is formed from lead peroxide; the electrolyte is dilute sulphuric acid. Lead acid cells now use a gel in place of a liquid for the electrolyte; therefore they don't have to be kept upright like wet cells.

Dry cells/batteries

Primary and secondary dry cells have a paste type electrolyte. Primary types are commonly used for many different applications, for example transistor radios, torches, PIRs, smoke detectors. Secondary types also have many different applications, for example, standby supplies, solar-powered security alarm sirens and digital cameras.

Battery capacity

The capacity of a battery (or a cell) is measured in ampere-hours (Ah). For example, if a battery delivers 10A for 10 hours, it has a capacity of 100Ah at the 10-hour rate; taking more than 10A will discharge the battery in less than 10 hours. Smaller batteries are rated in milliampere-hours (mAh), for example, a 3.7V lithium ion (Li-ion) 'rechargeable' battery for a digital camera is typically 740mAh and a 1.2V nickel-metal hydride (Ni-MH) 'rechargeable' battery for a PV garden light is typically 600mAh.

Figure 6.24 *Typical dry cell battery*

Remember

Battery electrolyte is very corrosive – avoid contact with the acid electrolyte. Do not create any electrical short circuits across the battery terminals.

Try this

A 6V lead-acid battery will have _____ cells and a 24V battery will have _____ cells.

Cells store electrical energy in a _____ form.

The three components in a simple cell are: a _____ plate, a _____ plate and an _____ .

Part 5 Transformers

Purpose of a transformer

The transformer is an extremely useful piece of electrical equipment. Its main use is to take an ac supply at one voltage and produce from this another ac supply. The voltage of the second supply may be quite different from that of the first and the process can isolate one from the other, a feature which makes the transformer ideal for the provision of safety services.

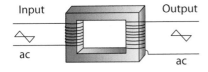

Figure 6.25 *Double-wound transformer*

Construction and enclosures

Small power transformers for use in equipment are usually of open construction, with varnished windings, and are air cooled.

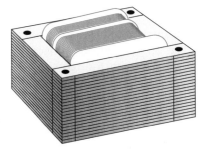

Figure 6.26 *Small air-cooled power transformer*

Larger transformers for medium power applications may be housed in metal tanks which are then filled with mineral oil which helps to insulate the windings. Excess heat is dissipated through the sides of the tank.

Large power transformers are enclosed in a metal tank with cooling tubes fitted to the outside through which the oil is free to circulate. The oil serves as an insulating and a cooling medium as the natural convection of the liquid carries away the heat from the windings as well as insulating one from the other.

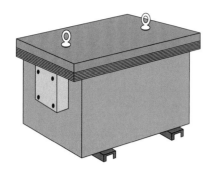

Figure 6.27 *Oil-cooled transformer*

Figure 6.28 *Large oil-filled transformer*

Operation of transformers

For all practical purposes, a transformer will consist of two coils of insulated wire wound around the same magnetic core.

One coil, which will be called the 'primary' will be connected to an ac supply. The alternating current in the primary coil will set up an alternating magnetic flux in the core and this will link with the turns of the 'secondary' winding. The alternating voltage thus produced will depend on the number of turns of wire in the secondary, just as the strength of the field will depend on the number of turns in the primary.

Assuming for the moment that there are to be no losses in the process then the ampere-turns in the primary winding will be equal to the ampere-turns in the secondary. Using the symbol N for the number of turns.

$$N_p I_p = N_s I_s \qquad \text{(equation 1)}$$

On the same theme, and assuming that the device has no losses then we can say that;

$$\text{Volt Amps in} = \text{Volt Amps out}$$

In other words

$$V_p I_p = V_s I_s \qquad \text{(equation 2)}$$

Going back to equation 1, this can be re-written as

$$\frac{N_p}{N_s} = \frac{I_s}{I_p}$$

and similarly, equation 2 can be written;

$$\frac{V_P}{V_S} = \frac{I_S}{I_P}$$

This gives us

$$\frac{V_P}{V_S} = \frac{N_P}{N_S} = \frac{I_S}{I_P}$$

This is often referred to as the basic transformer equation.

Step-down transformers

Transformers are widely used for electrical and electronic applications because they can change voltages from one level to another with relative ease.

If a transformer has 240 turns in the primary winding and 24 turns in the secondary it is said to have a transformer ratio of 10:1.

In other words the primary voltage is reduced by a factor of 10.

This relationship can be expressed by the basic equation

$$\frac{V_P}{V_S} = \frac{N_P}{N_S}$$

Thus a transformer with a 10 to 1 ratio will have a secondary voltage which is relative to the primary namely

$$V_S = V_P \times \frac{N_S}{N_P}$$

Taking the previous example;

If a transformer having 240 turns in the primary and 24 turns in the secondary is connected to a 125V supply the secondary voltage will be

$$V_S = V_P \times \frac{N_S}{N_P} = 125 \times \frac{24}{240} = 12.5V.$$

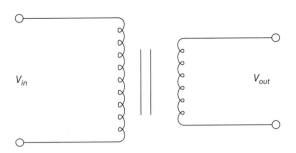

Figure 6.29 *Step-down transformer*

Step-up transformers

A transformer can just as easily be used for the purpose of raising the voltage and where higher voltages are required for a particular application the transformer provides a quick and easy solution.

For example if an item of equipment which is mainly operated at 24V ac but for one particular process requires 600V then the solution could be the inclusion of a transformer having a step up ratio of 25 to 1.

Given that the primary winding is to contain 50 turns then the secondary would need to be 25 times greater.

$$V_S = V_P \times \frac{N_S}{N_P} = 24 \times \frac{1250}{50} = 600V$$

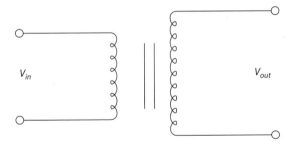

Figure 6.30 *Step-up transformer*

Example:

A single-phase step-down transformer has 760 primary turns and 360 secondary turns.

Calculate the secondary

a voltage if the primary voltage is 230V ac
b current if the primary current is 5A.

a $V_S = V_P \times \dfrac{N_S}{N_P} = 230 \times \dfrac{360}{760} = 109V$

b $I_S = I_P \times \dfrac{V_P}{V_S} = 5 \times \dfrac{230}{109} = 10.55A$

Volts per turn

Since both windings of a double-wound transformer are 'linked' by the same magnetic flux the induced emf per turn will be the same for both windings. Therefore, the emf in both windings is proportional to the number of turns.

Thus:

The volts per turn on the primary winding = the volts per turn on the secondary winding

or $\dfrac{V_P}{N_P} = \dfrac{V_S}{N_S}$.

Example:

A double-wound 230V/50V single-phase transformer has 110 primary turns. Calculate the volts/turn on the primary and secondary windings.

Volts per turn Primary $= \dfrac{V_P}{N_P} = \dfrac{230}{110} = 2.09$

$N_S = N_P \times \dfrac{V_S}{V_P} = 110 \times \dfrac{50}{230} = 23.9$ turns

Volts per turn Secondary $= \dfrac{V_S}{N_S} = \dfrac{50}{23.9} = 2.09$

Try this

A transformer is to be used to provide a 57.5V output from a 230V ac supply.

Calculate the:

1 turns ratio required _____

2 number of primary turns, if the secondary is wound with 500 turns _____

Try this

A transformer with 500 primary turns and 125 secondary turns is fed from a 230V ac supply. Calculate the:

1   secondary voltage

_____

_____

2   volts per turn

_____

_____

Try this

Small power transformers are _____ cooled. Large power transformers are enclosed in a metal _____ with _____ tubes fitted to the outside through which the oil _____.

In a transformer the _____ current in the _____ winding sets up an alternating magnetic _____ in the _____ and this links with the turns of the _____ winding to induce an _____ into it by means of _____ induction.

A step _____ transformer has more _____ turns and a step_____ transformer has more _____ turns.

The volts per turn on the primary = the _____ per turn on the _____.

Part 6 Isolating transformers

It is not always the case that transformers are used to change the voltage and for reasons of safety it may be necessary to provide a mains voltage supply which is not derived directly from the mains supply.

This is the principle adopted in the BS EN 61558-2-5 shaver socket in which the 230V shaver supply is provided by a socket outlet which has no reference connection to earth potential. It is therefore incapable of delivering an earth leakage current.

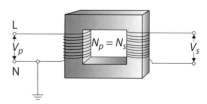

Figure 6.31 *Isolating transformer with no connection to earth on the secondary winding*

Voltage/current relationships

Because the transformer is a very efficient item of equipment then for most practical purposes its efficiency can be assumed to be very close to 100%.

That is to say, the power delivered to the primary winding is assumed to be equal to the power delivered by the secondary (power in = power out).

For example, a 5kVA transformer is supplied with 20A at 250V. Using this very basic relationship we could deduce that such a device would be capable of delivering 10A at 500V if the ratio happened to be 1:2.

Alternatively, if the secondary voltage happened to be 100V the current would have to be 50A in order to maintain the same primary current and thus the ratio would be a step down in the order of 2.5:1.

Transformer power ratings

Transformers are rated in VA, kVA or MVA (depending on their size); this is the transformer's maximum output power rating.

There are many different sizes available and choosing the correct size depends upon the total connected load on the secondary side. Some typical sizes are 15kVA, 30kVA, 45kVA, 75kVA and 150kVA.

Calculating the kVA rating of a single-phase transformer

$$kVA = \frac{V \times I}{1000}$$

Example:

Calculate the kVA rating of a transformer suitable to supply a 230V, 100A single-phase ac load.

$$kVA = \frac{V \times I}{1000} = \frac{230 \times 100}{1000} = \frac{23000}{1000} = 23kVA$$

Therefore a 30kVA transformer would be suitable.

Remember

Although transformers have no moving parts they are not 100% efficient.

Auto-transformers

Not all transformers are of the double-wound type (Figure 6.32). The auto-transformer has only one winding and is capable of stepping up and down the voltage as effectively as the double-wound variety.

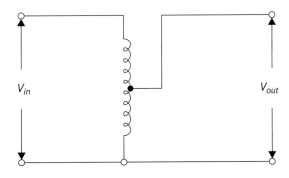

Figure 6.32 *Step-down auto-transformer*

Try this

Determine the kVA rating of a transformer suitable to supply a 230V, 60A single-phase ac load.

_____

_____

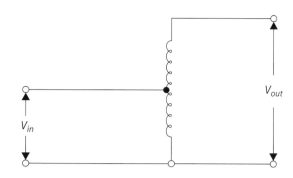

Figure 6.33 *Step-up auto-transformer*

One important advantage of the autotransformer is that the part of the winding which is common to both primary and secondary current carries the difference between the two currents.

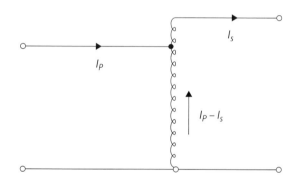

Figure 6.34 *Current in an auto-transformer*

Where this type of transformer is used to make small adjustments in mains voltage the currents are very nearly equal and therefore the resultant current in the common part of the winding is quite small. For this reason, the common section can be wound in comparatively light wire resulting in considerable cost and weight savings as compared with double-wound types.

The main problem with the auto-transformer arises from the fact that it does not have an isolated secondary. Another unfortunate feature is that if the common terminal should become disconnected then full input voltage will appear at the output terminals.

Remember

An auto-transformer has one winding and its operation relies on the principles of self-induction.

Three-phase transformers

Three-phase transformers are an essential feature of transmission and distribution systems.

The transformer ratio rules will still apply but with a three-phase transformer there is always the choice of star or delta connection.

The effect of this can be seen as follows.

Example:

The transformer has a ratio of 13:1 and is supplied at 22.52kV.

The primary windings are delta connected and the LV side is connected in star.

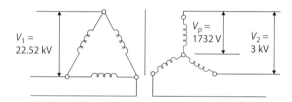

Figure 6.35 *Delta/Star transformer*

The voltage across each of the HV windings is 22.52kV and assuming negligible losses, each LV winding will have a voltage of

$$\frac{22.5 \times 10^3}{13} = 1732.3\text{V}$$

Since these windings are now to be connected in star connection the terminal voltage will be $V_2 \times \sqrt{3} = 1732.3 \times \sqrt{3} = 3\text{kV}$.

Try this

A three-phase transformer has a ratio of 27.5:1 and is supplied at 11kV. Determine the voltage across the secondary windings when they are connected in

1 **Delta**

_____

_____

2 **Star**

_____

_____

Part 7 Transformer efficiency

In practical terms it is clear that a transformer will not deliver exactly the same amount of power as it receives.

The efficiency of a transformer can be determined in a similar manner to efficiency calculations performed on any other form of energy-converting device.

$$\frac{\text{Output}}{\text{Input}} = \text{efficiency}$$

This is of course per unit efficiency and the result will be a fraction.

It is common practice to express efficiency as a percentage. This is nothing more than the per unit (pu) efficiency with the decimal point shifted two places to the right.

Example:

A double-wound transformer supplied with 16A at 180V supplies a load of 36A at a terminal voltage of 75.2V. What is the efficiency of the transformer?

$$\text{Efficiency} = \frac{\text{Power out}}{\text{Power in}} = \frac{36 \times 75.2}{16 \times 180} = \frac{270720}{2880} = 0.94\text{pu}$$

or alternatively

$$\% \text{ Efficiency} = \frac{\text{Power out} \times 100}{\text{Power in}} = \frac{270720}{2880} = 94\%$$

Transformer losses

A transformer is not 100% efficient because it has losses.

If one tenth of the power delivered to the transformer is taken up by the losses then only nine tenths of the input power would end up as output.

In other words

$$\text{Input} = \text{Output} + \text{Losses}$$

Transformer losses are generally grouped into two different categories.

● copper losses and
● iron losses.

Try this

Calculate the percentage efficiency of a double-wound 230V/110V transformer fed at 10A and supplying a load of 20A.

_____

_____

When transformer losses are taken into consideration the formula given below is used to calculate transformer efficiency.

$$\text{Efficiency} = \frac{\text{Output}}{\text{Output} + \text{losses}} \times 100\%$$

Example:

The full-load copper and iron losses of a transformer are 15kW and 10kW respectively. If the full-load output of the transformer is 900kW calculate the losses and the efficiency of the transformer on full load.

$$\text{Total loss} = \text{copper loss} + \text{iron loss}$$
$$= 15 + 10 = 25\text{kW}$$

$$\text{Efficiency} = \frac{\text{Output}}{\text{Output} + \text{Losses}} \times 100\%$$
$$= \frac{900}{900 + 25} \times 100 = 97.3\%$$

Instrument transformers

Current and voltage transformers are a common feature of fixed panel meter installations.

Current transformers

Current transformers, or CTs as they are commonly called, consist of a ring-shaped core around which is wound several turns of wire which make up the secondary winding. The primary winding is usually a single conductor such as a large cross-section cable or bus-bar.

Try this

The iron loss for a transformer is 6kW and its full-load copper loss is 9kW. If the full-load output is 500kW calculate the total losses and the efficiency at full load.

_____

_____

The secondary of an _____ transformer has no physical electrical connection to the _____ supply and no connection to _____ .

If a transformer is assumed to be 100% efficient power in = _____ .

An auto-transformer has only one _____ .

When losses are taken into consideration the formula for transformer efficiency is:

_____

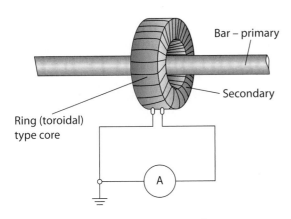

Figure 6.36 *Bar primary current transformer*

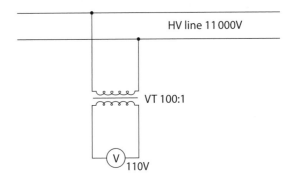

Figure 6.37 *Voltage transformer*

The secondary winding consists of a number of turns of insulated wire determined by the required output current.

For example if a current of 650A is to be indicated on a scale of 0 to 1000A using an ammeter whose full scale deflection is 5A then the CT would need to have an effective ratio of 1000:5 (or 200:1).

A current transformer must never be left in place on a loaded conductor without a secondary load of some description. A loaded primary conductor always produces a secondary voltage, and without a secondary load this can be high enough to damage the CT or even cause an accident.

If the ammeter has to be removed or disconnected for any reason always short-circuit the CT terminals first before disconnecting it. This will not harm the CT in any way and will prevent a dangerous situation from arising.

Voltage transformers

Voltage transformers, often abbreviated to VTs, are used to connect a voltmeter or voltage coil to a mains supply. The use of the VT is essential for high voltage measurement such as 11kV or 33kV, but even at 400V or 230V the isolation from a direct mains connection and a possible reduction to 110V can lead to increased safety.

A voltage transformer is a straightforward double-wound transformer of conventional construction and is not to be shorted out when not in use. The secondary terminals should, however, be insulated and enclosed to prevent accidental contact.

Measuring voltage, current and power on high voltage/current ac systems

Figure 6.38 shows the use of instrument transformers for metering purposes on high voltage and high current ac systems.

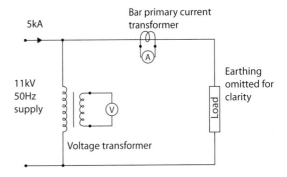

Figure 6.38 *Measuring voltage and current with CT and VT*

Figure 6.39 shows the connection of a single-phase wattmeter used to measure the power of a high voltage and high current ac system.

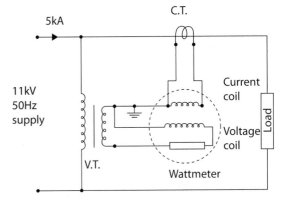

Figure 6.39 *Power measurement using CT and VT*

The current and voltage transformer effectively isolate the wattmeter from the high current/voltage system.

Clamp-on ammeter (tong tester)

The clamp-on ammeter is effectively a bar primary current transformer; the conductor under test is the bar primary and, as with most double-wound transformers, the secondary coil is wound on a former and fitted on a laminated iron core. This iron core is different to others insomuch as it is made to open up and allow room for a conductor to go into the centre.

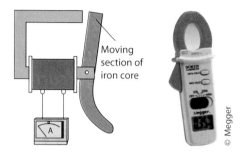

Figure 6.40 *Clamp meter for measuring current*

The moving section of laminated iron core is sprung so that when the gap is closed there is a tight joint between the two surfaces. The method of opening the core and where it opens depends on particular manufacturers.

Remember

The clamp-on ammeter allows current measurements to be taken without having to disconnect the supply and the circuit.

Electronic transformer

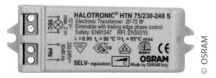

Figure 6.41 *Electronic transformer*

An electronic transformer is different in design to a conventional transformer with an iron core and windings, and is also much smaller and lighter. It has an inverter which conditions the current and voltage and changes the 50Hz supply frequency at the input to a high frequency output, typically 20–50kHz.

This type of transformer is widely used for providing an extra low voltage (ELV) 12V ac output for lighting circuits and many also have dimmable and soft-start control. Some types of electronic transformers use a rectifier to provide an ELV dc output.

Electronic transformers should be disconnected prior to carrying out insulation resistance tests on lighting circuits to prevent permanent damage to the transformers.

Try this: Crossword

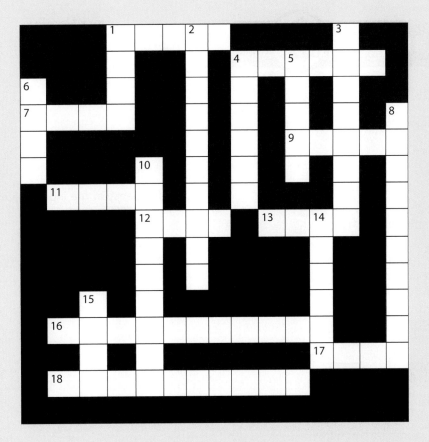

Across

1 Unit of length. (5)
4 Type of power. (6)
7 What inductive and capacitive reactance are measured in. (4)
9 Transformer star point is connected to this. (5)
11 Insulating material. (4)
12 Bar primary CT core shape. (4)
13 Winding. (4)
16 Often double wound. (11)
17 SI base unit. (4)
18 Drives a power station generator. (3, 7)

Down

1 Measured in kilograms. (4)
2 Opposes current flow. (10)
3 Factor that affects resistance. (8)
4 SI base unit. (6)
5 Square root of nine. (5)
6 Fossil fuel. (4)
8 Type of electrical supply. (5, 5)
10 Has a North and South Pole. (3, 6)
14 To turn a fraction upside down. (6)
15 Product of two lengths multiplied together. (4)

Congratulations you have now completed Chapter 6. Correctly complete the self-assessment questions before you progress to Chapter 7.

SELF ASSESSMENT

Circle the correct answers.

1 The power station generator output voltage is normally:

  a. 11kV
  b. 33kV
  c. 25kV
  d. 132kV

2 In Figure 6.42 the cells are connected:

  a. x series and y series
  b. x parallel and y series
  c. x series and y parallel
  d. x parallel and y parallel.

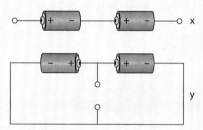

Figure 6.42 *Multiple cells*

3 A step-down transformer with 1 000 primary turns and 250 secondary turns is supplied at 250V ac. The secondary voltage will be:

  a. 230V
  b. 62.5V
  c. 25V
  d. 125V.

4 Which of the following is the cheapest method of producing electricity?

  a. gas
  b. oil
  c. coal
  d. hydro-electric.

5 If the full-load output of a transformer is 750kW and its total losses are 20kW, the transformers efficiency at full-load will be:

  a. 57%
  b. 97.4%
  c. 67%
  d. 82.4%.

Progress check

1. The sum of $8\frac{3}{4} + 2\frac{5}{8}$ is:

   ☐ a. $14\frac{1}{8}$

   ☐ b. $14\frac{3}{8}$

   ☐ c. $8\frac{1}{4}$

   ☐ d. $11\frac{3}{8}$

2. The value of y in the equation 8y + 12 = 36 is:

   ☐ a. 2

   ☐ b. 3

   ☐ c. 4

   ☐ d. 6

3. If the tangent of an angle is 0.5, the angle will be:

   ☐ a. 60°

   ☐ b. 26.57°

   ☐ c. 30°

   ☐ d. 86.57°

4. Mass is measured in:

   ☐ a. kilograms

   ☐ b. newtons

   ☐ c. joules

   ☐ d. coulombs

5. What is the inductive reactance of a coil which has a self inductance of 0.636H and negligible resistance when connected to a 230V 50H$_z$ supply?

   ☐ a. 63.6Ω

   ☐ b. 100Ω

   ☐ c. 199.8Ω

   ☐ d. 919.2Ω

6. In the circuit shown the product of the voltmeter and ammeter readings will give:

   ☐ a. true power

   ☐ b. actual power

   ☐ c. power factor

   ☐ d. apparent power

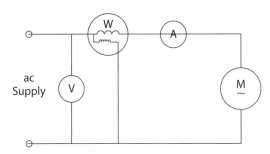

7. A storage heater element is rated at 240V, 840W. Neglecting the temperature of the element what is its resistance?

   ☐ a. 0.285Ω

   ☐ b. 3.5Ω

   ☐ c. 12.25Ω

   ☐ d. 68.5Ω

8. The kinetic energy of an object with a mass of 400kg and a velocity of 10m/s is:

   ☐ a. 10 000J

   ☐ b. 20 000J

   ☐ c. 30 000J

   ☐ d. 40 000J

9. The output power of a motor is 9 kW. If the percentage efficiency of the motor is 75%, what was the input power?

   ☐ a. 6.75kW

   ☐ b. 10.25kW

   ☐ c. 12KW

   ☐ d. 15kW

10. **Copper is easy to draw out into wires because it is very:**

☐ a. rigid

☐ b. hard

☐ c. ductile

☐ d. flexible

11. **What is the equivalent resistance of a 12Ω resistor connected in parallel with a 6Ω resistor?**

☐ a. 4Ω

☐ b. 6Ω

☐ c. 18Ω

☐ d. 72Ω

12. **The total power dissipated by two 230V tungsten lamps of resistance 529Ω each when connected in parallel to a 230V supply is:**

☐ a. 120W

☐ b. 200W

☐ c. 80W

☐ d. 40W

13. **When applying Fleming's left-hand rule, the first finger indicates the direction of the:**

☐ a. field

☐ b. induced current

☐ c. motion

☐ d. induced emf

14. **The current carrying conductor in the diagram will move:**

☐ a. downwards

☐ b. upwards

☐ c. to the left

☐ d. to the right

15. **The rms voltage is found to be 210V. This means that the peak voltage must be approximately:**

☐ a. 105V

☐ b. 210V

☐ c. 297V

☐ d. 330V

16. **The star point of a transformer winding is connected to:**

☐ a. the delta point and earth

☐ b. two phases and earth

☐ c. grey phase and neutral

☐ d. earth and neutral

17. **Super grid transmission voltage is:**

☐ a. 400kV

☐ b. 132kV

☐ c. 33kV

☐ d. 25kV

18. **A transformer has 150 turns on the input side and 600 turns on the output side. If the output voltage is 200 V, what is the input voltage?**

☐ a. 800V

☐ b. 400V

☐ c. 100V

☐ d. 50V

19. **What effort would be required to lift a load of 10kg placed 600mm from the fulcrum of a lever? The distance between the fulcrum and the point the force is to be exerted is 750mm.**

☐ a. 8N

☐ b. 23N

☐ c. 78.48N

☐ d. 784.8N

20. **One limitation of a macro wind turbine is:**

☐ a. it's renewable

☐ b. it has an impact on the landscape

☐ c. it's ideal for remote locations

☐ d. it has comparatively low maintenance

Ac circuits and systems

RECAP

Before you start work on this chapter, complete the exercise below to ensure that you remember what you learned earlier.

1 The 11kV input to a local substation transformer is connected in _____ and the 400/230V output is connected in _____.

2 The TT earth fault loop path comprises the:

- circuit _____ conductor

- consumer's earth _____ and _____ conductor

- return path, the mass of _____ and the supply earth _____

- supply transformers earthed _____

- transformer _____

- _____ conductor from the transformer to the fault.

3 Sketch the winding arrangements for *each* of the following types of single-phase transformer.

   a) step-down auto-transformer   **b)** step-up auto-transformer

4 Measurements have been taken on the secondary side of a single-phase transformer. The voltmeter reading is 240V and the ammeter reading is 80A. Using these readings, calculate the kVA output of the transformer.

_____

_____

LEARNING OBJECTIVES

On completion of this chapter you should be able to:

● Explain the relationship between resistance, inductance, capacitance and impedance in ac circuits.

● Calculate unknown values of resistance, inductance, inductive reactance, capacitance, capacitive reactance and impedance.

● Explain the relationship between kW, kVA, kVA_r and power factor.

● Calculate power factor.

● Explain power factor correction and ways of improving power factor.

● Explain load balancing and determine the neutral current in a three-phase and neutral supply.

● Calculate values of voltage and current in star and delta systems.

Part 1 Resistance, inductance, capacitance and impedance in ac circuits

We have already covered resistance, inductance, capacitance and impedance in previous chapters. To refresh your understanding of these commodities Part 1 is essentially a revision exercise of information we have already learned.

In an R–L–C series circuit: R, X_L, X_C and Z can be found using the following formulae:

$$I = \frac{V}{R} \text{ and likewise } X_L = \frac{V_L}{I} \text{ and } X_C = \frac{V_C}{I} \text{ and also } Z = \frac{V_T}{I}$$

also: $X_L = 2\pi fL$ and so $L = \dfrac{X_L}{2\pi f}$ and

$$X_C = \frac{1}{2\pi fC} \text{ and so } C = \frac{1}{2\pi fX_C} \text{ or } \frac{10^6}{2\pi fX_C} \text{ when C is in } \mu F$$

Angles may be measured in degrees or radians. In the formulae opposite 2π represents one complete ac cycle (each second) and 1 cycle = 1 revolution = 360° = 2π radians. 1 radian = 57.3° to 1 decimal place.

Therefore: $2 \times \pi \times 57.3 = 360° = 1$ cycle.

Remember

The number of complete cycles occurring each second is:

50 cycles per second (cps) = 50 hertz (Hz).

Try this

In an ac circuit; the _____ (R), inductive _____ (X_L) and _____ reactance (X_C) oppose the flow of alternating _____. The total _____ to the flow of alternating current is the _____ (Z) of the circuit, which is the effective opposition to alternating current flow of all the components (resistance, inductance and capacitance) in the ac circuit. Since R, X_L, X_C, and Z all oppose alternating current flow they are all measured in _____.

Remember the formula for Ohms law: $I = \dfrac{V}{R}$ and so $R = \dfrac{V}{I}$.

Try this

Using appropriate formulae, complete the table by inserting the missing values if the frequency in all cases is 50Hz. Cells highlighted in grey can be ignored.

R	X_L	X_C	Z	V	I	L	C
				230V		2H	
				230V			10µF
	100Ω				2.4A		
300Ω		398Ω					
	40Ω		50Ω	220V			
30Ω	100Ω	60Ω			0.48A		

Part 2 Effects of resistance, inductance, capacitance and impedance in ac circuits

Having completed the revision exercise let's now look at the effects that these four electrical variables have on two other electrical variables (alternating current and voltage) in ac circuits. It is more convenient to use phasor diagrams (also referred to as vector diagrams in other disciplines) instead of waveform diagrams to represent ac variables as they give a clearer picture of the different phase angles.

Phasor diagram representation of ac variables

The magnitude (size) of the voltage or current is represented by drawing a straight line to scale or by writing the magnitude of the voltage or current at the side of the line. The direction of the voltage or current is indicated by the position of the line on the diagram.

Example:

When an alternating voltage is applied to a circuit it is found that the current produced lags the voltage by 45°. Construct a wave and a phasor diagram for this circuit.

Since the magnitude of the voltage and current are not given it is not necessary to draw the diagrams to scale.

There are two possible answers both equally correct.

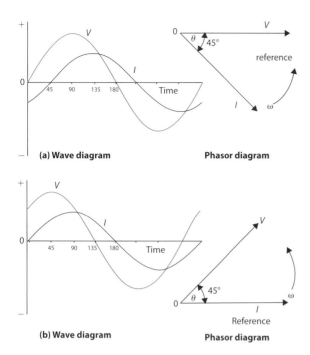

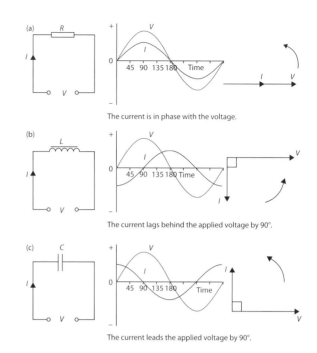

Figure 7.1 *Wave and phasor diagrams*

Figure 7.2 *Relationship between current and applied voltage*

The conventional direction of rotation of a phasor is anti-clockwise therefore the current lags behind the voltage by 45° in both diagrams (a) and (b).

So which do we choose?

a Phasor diagram (a) with V as the reference phasor is suitable for a parallel circuit since the supply voltage is the same for each branch in a parallel circuit.

b Phasor diagram (b) with I as the reference phasor is suitable for a series circuit since the same current flows in each part of the circuit.

Figure 7.2 shows how phasor diagrams can be used to represent the relationship between current and applied voltage in a:

a purely resistive circuit
b purely inductive circuit
c purely capacitive circuit.

and the applied voltage (V) is taken as the reference.

Note

A useful memory aid is C I V I L

In a capacitive circuit the current leads the voltage.
In an inductive circuit the current lags the voltage.

Phasor addition

Alternating variables cannot be added arithmetically when they are out of phase with each other.

Let's consider the simple R-L series circuit.

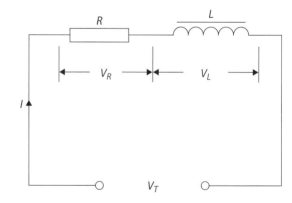

Figure 7.3 *Simple R-L series circuit*

The voltage across the resistor V_R cannot be added to the voltage across the inductor V_L to find the total voltage V_T because they are out of phase with each other.

Example:

In Figure 7.3 if the voltage across the resistor is 40V and the voltage across the inductor is 60V find the value of the supply voltage by drawing a phasor diagram to a scale of 1cm = 20V.

Answer:

Remember, *I* is the reference phasor in a series circuit.

Step 1. Draw the current phasor horizontally as 'reference' (not to any scale but to a suitable length).

Figure 7.4 *Current reference*

Step 2. Draw the phasors for V_R and V_L to scale.

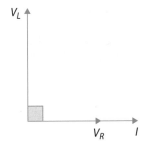

Figure 7.5 *Voltages added to the reference phasor*

Step 3. Construct the phasor parallelogram and hence determine the supply voltage V_T. This is represented by the diagonal line of the parallelogram.

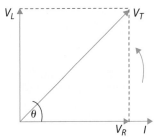

Figure 7.6 *Completed phasor showing V_T*

Measure the length of the diagonal line and you will find that V_T = 72V, which is the 'phasor sum' of V_L and V_R.

The angle θ is the number of degrees that *I lags* behind the supply voltage V_T and in this example the angle $\theta = 56°$.

Check this with your protractor.

We can also check the answer by applying Pythagoras' theorem.

$$V_T = \sqrt{V_R^2 + V_L^2} = \sqrt{40^2 + 60^2} = 72.11\,V\,(72\,V)$$

Try this

Apply Pythagoras' theorem to determine the voltage across the inductance V_L.

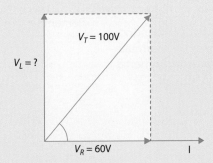

Figure 7.7 *Voltage triangle*

Part 3 Ac circuit calculations

R-L series circuits

Example:

A coil has a resistance of 6Ω and an inductance of 25.5mH. If the current flowing in the coil is 10A when connected to a 50Hz supply, determine the supply voltage by drawing a phasor diagram to a scale of 1cm = 10V.

(A coil is represented as an R-L series circuit.)

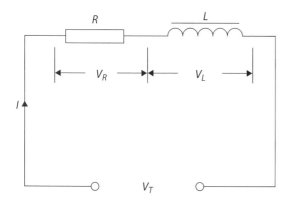

Figure 7.8 *Circuit for a coil*

Step 1. Calculate the values of V_R and V_L.

$$V_R = I \times R$$
$$= 10 \times 6$$
$$= 60\,V$$
$$V_L = I \times X_L$$
$$= 10 \times ?$$

However we cannot do this until we have found the inductive reactance X_L and to do this we use the formula: $X_L = 2\pi f\, L$ where f is the frequency of the supply (hertz) and L is the inductance of the coil (henrys).

$$= 2 \times 3.142 \times 50 \times 25.5 \times 10^{-3}$$
$$= 8\Omega$$

$$V_L = I \times X_L = 10 \times 8 = 80V$$

Step 2. Construct the phasor diagram and measure the diagonal line to find the answer.

Phasor diagram scale 1cm = 10V

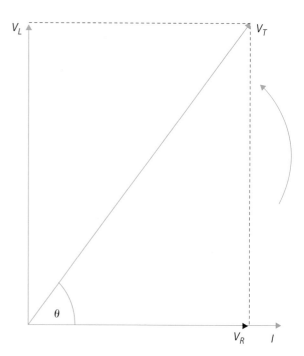

Figure 7.9 *Phasor diagram Answer = 100V (10 cm)*

Impedance – symbol Z

It should be seen from the previous example that there are two separate oppositions to the flow of current in an R-L circuit, one is due to the resistance (R) and the other is due to the inductive reactance (X_L). The combination of these oppositions is called the 'impedance' of the circuit and it is measured in ohms (Ω).

Remember the impedance triangle.

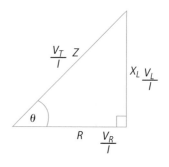

Figure 7.10 *Impedance triangle for R-L series circuit*

From the diagram:

$$Z = \frac{V_T}{I}, \; X_L = \frac{V_L}{I} \text{ and } R = \frac{V_R}{I}$$

$$\therefore V_T = IZ, \; V_L = IX_L \text{ and } V_R = IR$$

$$\therefore I = \frac{V_T}{Z}, \; I = \frac{V_L}{X_L} \text{ and } I = \frac{V_R}{R}$$

Remember that if the resistance and inductive reactance are known the impedance can be found by the formula $Z = \sqrt{R^2 + X_L^2}$.

R-C series circuit

Example:

A capacitor of 160µF is connected in series with a non-inductive resistor of 15Ω across a 50Hz supply. If the current drawn is 10A:

a calculate:

   (i) the capacitive reactance

   (ii) the voltage across each component

b find by means of a phasor diagram the supply voltage.

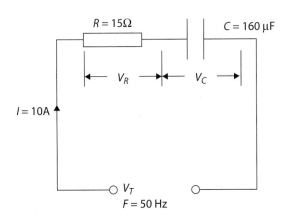

Figure 7.11 *R-C series circuit*

a (i) $X_C = \dfrac{10^6}{2\pi fC} = \dfrac{10^6}{2 \times 3.142 \times 50 \times 160} = 19.89\,\Omega$

   Say 20Ω

  (ii) $V_R = I \times R = 10 \times 15 = 150V$

    $V_C = I \times X_C = 10 \times 20 = 200V$

b
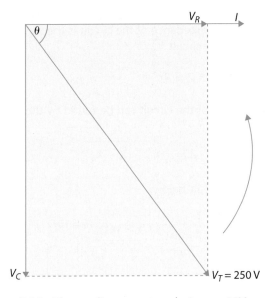

Figure 7.12 *Phasor diagram at scale 1cm = 30V*

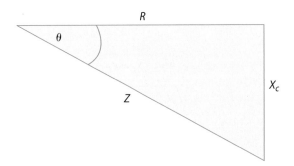

Figure 7.13 *Impedance triangle for R-C series circuit*

R-L-C series circuits

In the R-L-C series circuit there are three oppositions to the flow of current: resistance, inductive reactance and capacitive reactance.

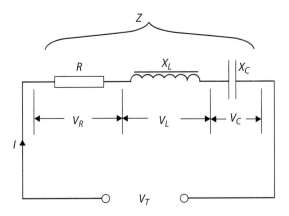

Figure 7.14 *Three oppositions to the flow of current*

If the inductive reactance is the same as capacitive reactance, then the impedance of the circuit is the same as the resistance of the circuit.

i.e. if $X_L = X_C$ then $Z = R$

The impedance of the circuit can be found by the formula

$$Z = \sqrt{R^2 + (X_L - X_C)^2}$$

Consider the circuit below.

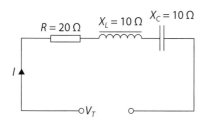

Figure 7.15 *R-L-C series*

$$
\begin{aligned}
Z &= \sqrt{R^2 + (X_L - X_C)^2} \\
&= \sqrt{20^2 + (10 - 10)^2} \\
&= \sqrt{20^2} \\
&= \sqrt{400} \\
&= 20\Omega
\end{aligned}
$$

We have proved that $Z = R$ when $X_L = X_C$.

Remember: power factor $= \dfrac{R}{Z}$ and when $Z = R$ the power factor of the circuit is 1. Therefore the power factor of the circuit in Figure 7.15 $= \dfrac{R}{Z} = \dfrac{20}{20} = 1$. (Power factor is covered in more detail later in this chapter.)

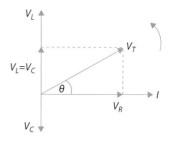

Figure 7.16 *Phasor diagram for R-L-C series circuit (assuming that V_L is greater than V_C)*

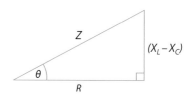

Figure 7.17 *Associated impedance triangle*

Try this

A pure inductor of 0.05H is connected in series with a non-inductive resistor of 25Ω to a 230V, 50Hz ac supply.

Calculate:

1   the impedance of the circuit _____

2   the current in the circuit _____

3   the voltage across (i) the inductor and (ii) the resistor_____

_____

I is used as the reference on a phasor diagram for a _____ circuit since the _____ current flows in each part of the circuit.

V is used as the reference on a phasor diagram for a _____ circuit since the supply _____ is the same for each branch of the circuit.

In a capacitive circuit the current _____ the voltage, and in an inductive circuit the current _____ the voltage.

Part 4 Power factor and its relationship with kW, kVA and kVA$_r$

In Chapter 2 we determined that power factor is the ratio of actual or true power to apparent power and it can be calculated by using the formula:

$$\text{power factor} = \cos \theta = \frac{kW}{kVA}$$

Remember the power triangle.

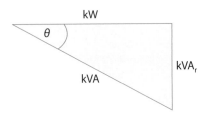

Figure 7.18 *Power triangle – when PF is lagging*

In the power triangle:

kW is the true or actual power

kVA is the apparent power

kVA$_r$ is the reactive power.

The R-L series circuit, as in Figure 7.8, can represent a motor or a transformer circuit, R being the resistance of their windings and L being the inductance of their windings, and it is due to this inductance that the power factor will be lagging.

The true (actual) power (kW) is dissipated in the circuit resistance (R) only. The reactive power (kVA$_r$) is the wattless power of the circuit inductance (L) only. The combination of the two is known as the apparent power (kVA).

From your answers to the Try this questions below, it should be clear that both the apparent power (kVA) and the reactive power (kVA$_r$) have both reduced in value when the power factor is improved. You should also see that the true power (kW) dissipated by the motor does not change.

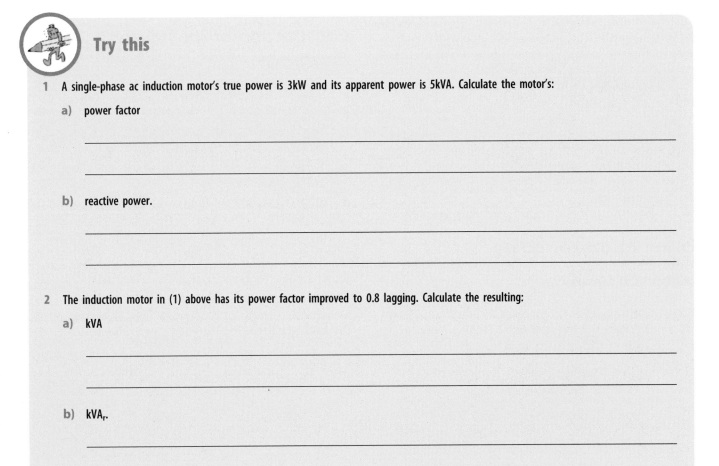

Try this

1 A single-phase ac induction motor's true power is 3kW and its apparent power is 5kVA. Calculate the motor's:

a) power factor

_____

_____

b) reactive power.

_____

_____

2 The induction motor in (1) above has its power factor improved to 0.8 lagging. Calculate the resulting:

a) kVA

_____

_____

b) kVA$_r$.

_____

_____

Part 5 Power factor improvement

Why improve the power factor?

Electrical energy supplied at a low power factor is costly to the supply authority since the larger currents taken from the supply require heavier cables and switchgear than is really necessary.

The industrial consumer also pays more than the cost of the true power used because their kilovolt ammeter (kVA meter) records the instantaneous values of the product of voltage and current.

Therefore it is in the interests of both supply authority and consumer to keep the power factor as near as possible to unity.

Methods of power factor improvement/correction

The most common method used for power factor improvement is to use capacitors. PF improvement capacitors are often fitted to individual items of electrical plant (such as motors and fluorescent luminaires). Banks of capacitors may be connected to the supply intake terminals to improve the power factor of the whole electrical installation and plant. However, this method needs automatic variable control as the plant is switched on and off.

Synchronous motors are often installed to drive constant loads, such as pumps or fans, but are also occasionally used for power factor improvement and correction. (This method is covered in more detail in the next chapter.)

Power factor correction

Correction to unity

A 230V, 50Hz induction motor takes a current of 50A and has a power factor at this load of 0.8 lagging. We need to find the value of the capacitor to correct the power factor to unity.

First it will help to draw a simple circuit and to put all the known values on the diagram.

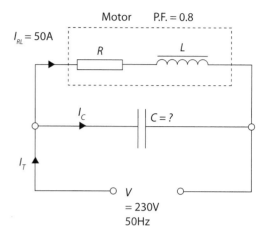

Figure 7.19 *The motor circuit*

We can then provide a graphical solution, by constructing a phasor diagram to scale.

Construction of the phasor diagram

a   Draw the horizontal reference V (not to scale).

b   Find the angle θ.
    0.8 cos^{-1} = 36.87° on your calculator (on some calculators you will need to press 0.8 INV cos)

c I_{RL} can now be drawn to scale at the angle θ using a ruler and protractor.

d   (i)   Draw a vertical line to meet the reference and label it I_L.

      (ii)   Draw another vertical line at 90° to V the same length as (i) and label it I_C.

e   The parallelogram may now be completed.

Measure the length of the vertical line which represents the leading current the capacitor must produce.

I_C = 30 Amps

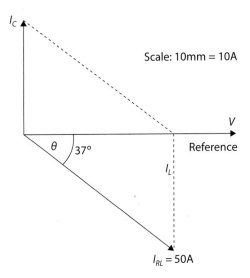

Scale: 10mm = 10A

Figure 7.20 *Phasor diagram for power factor correction*

Next find the capacitive reactance (X_C) of the capacitor.

$$X_C = \frac{V}{I_C} = \frac{230}{30} = 7.66\Omega$$

Since

$$X_C = \frac{10^6}{2\pi f C}$$

the value of the capacitor (in microfarads) can now be found by transposing the formula for C.

(Note: f is the supply frequency)

$$C = \frac{10^6}{2\pi f X_C}$$

$$= \frac{10^6}{2 \times 3.142 \times 50 \times 7.66}$$

$$= 415.5\,\mu F$$

Try this

The phasor diagram below represents to scale the current taken by a 230V, 50Hz motor. A capacitor is connected across the terminals of the motor to raise the power factor to unity.

1   Use the diagram to find the current taken by the capacitor.

_____

_____

2   Calculate the capacitance of the capacitor.

_____

_____

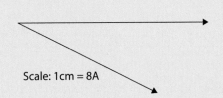

Scale: 1cm = 8A

Figure 7.21 *Phasor diagram*

Single-phase power

To calculate the power in a single-phase ac circuit which contains resistance and inductance we use:

$$Cos\theta = \frac{\text{True power}}{\text{Apparent power}} \text{ and therefore}$$

True power = Apparent power × Cosθ and since Apparent power = V × I (VA) then

$$\text{True power} = V \times I \times Cos\theta \text{ (Watts)}.$$

So the power in a single-phase ac circuit can be calculated by $P = VI \cos \theta$, where θ is the angle by which the current and voltage are out of phase.

We can also calculate the circuit current using:

$$I = \frac{P}{V \times Cos\,\theta}$$

Try this

Calculate the current drawn by a 230V, 3kW, single-phase load with a power factor of 0.8 lagging.

_____

_____

Power factor can be improved by connecting a _____ across the _____ in parallel with an inductive load.

When power factor is poor the supply current is _____ and cables and switchgear need to be _____.

The true power of a single-phase ac circuit containing resistance and inductance can be found using the formula _____

and the circuit current can be found by transposing the formula to _____.

Part 6

Three-phase circuit calculations

Voltage and current in star and delta connected systems

Remember this diagram from Chapter 6.

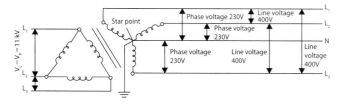

Figure 7.22 _Delta/Star connected transformer_

The line voltage V_L is the same value as the phase voltage V_P across each delta connected primary winding.

The line voltage V_L is different to the phase voltage V_P on the star connected secondary side.

The relationship between the phase voltage and the line voltage for a star connected winding is:

$$\text{Phase voltage } (V_P) = \frac{\text{Line voltage } (V_L)}{\sqrt{3}} \text{ and}$$

$$\text{Line voltage } (V_L) = \text{Phase voltage } (V_P) \times \sqrt{3}$$

Example:

The phase voltage is 230V for most domestic premises. The line voltage for this supply is $V_L = 230 \times \sqrt{3} = 230 \times 1.73 = 400V$

Similarly a 400V line voltage gives a phase voltage of $V_P = \dfrac{400}{1.73} = 230V$.

Try this

For star connected windings

1 Calculate the phase voltages if the line voltages are:

  a) 240

  _____

  _____

  b) 425

  _____

  _____

  c) 300

  _____

  _____

2 Calculate the line voltages if the phase voltages are:

  a) 415

  _____

  _____

  b) 230

  _____

  _____

  c) 110

  _____

  _____

Load currents in three-phase circuits

It is important to recognize the relationships of the currents in star and delta connected windings. In star connected the current through the line conductors is equal to that flowing through the phase windings, as shown in Figure 7.23.

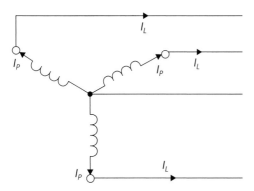

Figure 7.23 *Current in a system connected in star*

However in the delta connected winding this appears to be more complex. The line current, when reaching the transformer winding, is split into two directions so that two phase windings are each taking some current. As each of the phases are 120° out of phase with each other and the current is alternating each line conductor acts as a flow and return.

The arrows shown in Figure 7.24 only give an indication as to the current distribution from each line and all of these currents would not be flowing in the directions shown at the same time.

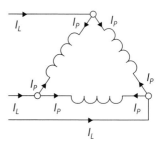

Figure 7.24 *Current in a system connected in delta*

The current through the phase windings are: $I_P = \dfrac{I_L}{\sqrt{3}}$ and $I_L = I_P \times \sqrt{3}$

Example:

If the line current is 100A the phase current is $I_P = \dfrac{I_L}{\sqrt{3}} = \dfrac{100}{1.73} = 57.8$ A.

Remember

For delta : $V_L = V_P$

$$I_L = \sqrt{3} \times I_P \quad \text{or} \quad I_P = \frac{I_L}{\sqrt{3}}$$

For star : $I_L = I_P$

$$V_L = \sqrt{3} \times V_P \quad \text{or} \quad V_P = \frac{V_L}{\sqrt{3}}$$

Try this

For delta connected windings

1   Calculate the phase currents for delta connected windings if the line currents are:

a)   100A

_____

_____

b)   240A

_____

_____

   c)   2 000A

___

2 Calculate the line currents if the phase currents are:

   a)   50A

___

   b)   240A

___

   c)   1kA

___

Three-phase power

In a three-phase system the power in each phase is the product of the phase voltage, phase current and power factor.

i.e. $V_P \times I_P \times \cos \theta$

therefore in a balanced system:

Total power $= 3 \times V_P \times I_P \times \cos \theta$

Note

Three-phase motors normally provide a balanced load.

Star connection

Power $= 3 \times V_P \times I_P \times \cos \theta$

But $V_P = \dfrac{V_L}{\sqrt{3}}$ and $I_P = I_L$

∴ Three-phase power $= \dfrac{3V_L I_L \cos \theta}{\sqrt{3}}$

$= \sqrt{3}\, V_L I_L \cos \theta$

Delta connection

Power $= 3 V_P I_P \cos \theta$

But $V_P = V_L$ and $I_P = \dfrac{I_L}{\sqrt{3}}$

∴ Three-phase power $= \dfrac{3V_L I_L \cos \theta}{\sqrt{3}}$

$= \sqrt{3}\, V_L I_L \cos \theta$

Remember

Three-phase power of a 'balanced system' (star or delta connected) is calculated by the formula:
Three-phase power $= \sqrt{3}\, V_L I_L\ \cos\theta$

Example:

Three identical 20 Ω impedances, each with a power factor of 0.85 lagging, are connected to a 400V three-phase supply. Calculate the power dissipated when the impedances are connected in:

a star
b delta

a
$$V_P = \frac{V_L}{\sqrt{3}}$$
$$= \frac{400}{1.732}$$
$$= 231V$$
$$I_L = I_P = \frac{V_P}{Z_P} = \frac{231}{20} = 11.55A$$
$$P = \sqrt{3}\, V_L I_L\ \cos\theta$$
$$= 1.732 \times 400 \times 11.55 \times 0.85$$
$$= 6801.564W\ (6.8kW)$$
OR total power $= 3 \times V_P \times I_P \times \cos\theta$
$$= 3 \times 231 \times 11.55 \times 0.85$$
$$= 6803.5275W\ (6.8kW)$$

b
$$I_P = \frac{V_P}{Z_P} = \frac{400}{20}\ (\text{since } V_P = V_L)$$
$$= 20A$$
$$I_L = \sqrt{3}\, I_P$$
$$= 1.732 \times 20 = 34.64A$$
$$P = \sqrt{3}\, V_L I_L\ \cos\theta$$
$$= 1.732 \times 400 \times 34.64 \times 0.85$$
$$= 20398.8W\ (20.4\,kW)$$
Or total power $= 3 \times V_P \times I_P \times \cos\theta$
$$= 3 \times 400 \times 20 \times 0.85$$
$$= 20400W\ (20.4kW)$$

Note

The power dissipated in delta is three times the power dissipated in star. (i.e. $3 \times 6.8 = 20.4kW$).

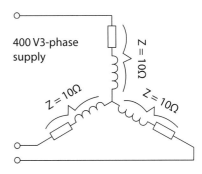

Figure 7.25 *Three-phase power*

Try this

1 If the impedances shown in Figure 7.25 each have a power factor of 0.9, calculate the total power dissipated.

_____

_____

2 If the same impedances are now connected in delta, determine the total power dissipation.

_____

_____

Example:

Three 30Ω resistors are connected in star to a 400V, three-phase, 4-wire ac supply.

Calculate:

a   the phase and line currents
b   the total power dissipated.

a $V_P = \dfrac{V_L}{\sqrt{3}} = \dfrac{400}{1.732} = 231V$

   $I_P = \dfrac{V_P}{R_P} = \dfrac{231}{30} = 7.7A$

In star $I_L = I_P = 7.7\ A$

b $P = \sqrt{3}V_L I_L \cos\theta$

   $= 1.732 \times 400 \times 7.7 \times 1$

   $= 5334.56W\ (5.3kW)$

Or total power $= 3 \times V_P I_P \cos\theta$

   $= 3 \times 231 \times 7.7 \times 1$

   $= 5336.1W\ (5.3kW)$

Remember for purely resistive ac loads the power factor is unity, therefore $\cos\theta = 1$.

Try this

For the circuit shown determine:

1   the line current

_____

_____

2   the phase current

_____

_____

3   the total power dissipated

_____

_____

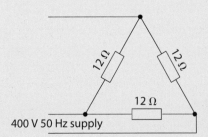

400 V 50 Hz supply

Figure 7.26 *Line and phase current and power dissipated*

A 30kW, 400V balanced three-phase delta connected load has a power factor of 0.86 lagging. Calculate (a) the line current (b) the phase current.

Part 7

Three-phase balanced loads

All transmission and distribution is carried out using a three-phase system. It is important that each of the phases carries about the same amount of current.

Three-phase motors have windings where each phase is the same and therefore the conductors carry the same current. These automatically create a balanced situation.

For domestic supplies the output of the star connected transformer is 400/230V. All premises are supplied with a phase and neutral of 230V. To try and balance the loads on each of the phases houses may be connected equally distributed across the three phases.

Remember

It is always important to try and balance the single phase loads across the phases of a three-phase supply.

If it was possible to load all of the phases exactly the same the current in the neutral would be zero.

This can be proved by drawing the loads to scale on a phasor diagram, as shown in Figure 7.27.

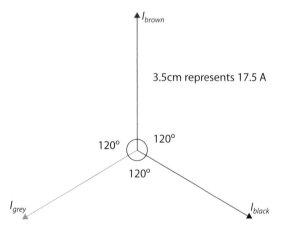

Figure 7.27 *Phase currents in a balanced three-phase system Scale: 1cm represents 5A*

Each line is drawn the same length to a scale equalling the current and each line is at equal angles (120°) to the one before, going in a clockwise direction.

By now 'adding' the load of each phase to the other in the direction of the arrow and at the correct angle the resultant current I_N can be found.

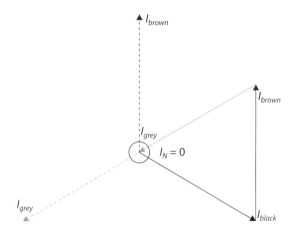

Figure 7.28 *Phasor sum of balanced load currents. Scale: 1cm represents 5A*

If the current in the phases are not equal then the resultant I_N would not be zero.

$I_{brown} = 150A$

$I_{black} = 200A$

$I_{grey} = 100A$

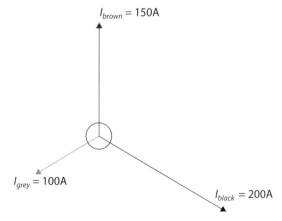

Figure 7.29 *Unbalanced three-phase load currents Scale: 1cm represents 50A*

The current in the neutral I_N can be found by measuring the distance from the star point to I_{grey}.

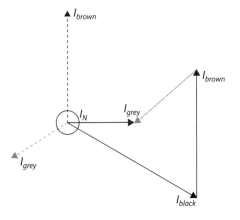

Figure 7.30 *Phasor sum of unbalanced load currents. Scale: 1cm represents 50A*

In this case it is

$I_N = 1.7$cm

$I_N = 85$A

This means that due to the phases not being balanced 85A is flowing in the neutral conductor.

Reasons for balancing single-phase loads on a three-phase system:

● to limit the current in the neutral conductor (I_N)

● to prevent excessive loading on any one phase conductor (brown, black or grey), thus keeping the size of the cable used to a minimum

● to reduce the voltage drop in the system (if $I_N = 0$ then $I_N \times R_N$ will be zero)

● to reduce the power loss in the system (if $I_N = 0$ then $I_N^2 \times R_N$ will be zero)

Try this

The currents on a three-phase star connected load are:

$I_{brown} = 25$A

$I_{black} = 15$A

$I_{grey} = 10$A

Draw a scaled phasor diagram at 1cm = 5A and determine the current in the neutral.

Three-phase power of a 'balanced system' (_____ or _____ connected) can be calculated by the formula:

_____

Single-phase loads should be balanced on a three-phase system mainly:

● to _____ the current in the _____ conductor

● to prevent any one phase taking excessive _____ current.

On a balanced three-phase system the _____ in the neutral would be _____ .

Congratulations you have now completed Chapter 7. Correctly complete the self assessment questions before you progress to Chapter 8.

SELF ASSESSMENT

Circle the correct answers.

1

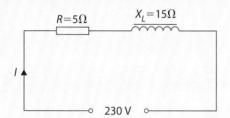

In the diagram the impedance of the circuit is:

a. 4.47Ω
b. 14.14Ω
c. 20Ω
d. 15.8Ω.

2   The power consumed by an inductor can be calculated from:

a. $P = VI$
b. $P = VI \cos \theta$
c. $P = \dfrac{V}{I \cos \theta}$
d. $P = \dfrac{V}{I}$

3   A single phase motor is connected so that its voltage, current and wattage can be monitored. One set of readings gives:

$V = 240V$, $I = 4.8A$ and $P = 920W$.

The power factor in this case is:

a.   1.2
b.   18.4
c.   0.8
d.   0.5.

4   The kVA input of a motor taking 8kW and 6kVA$_r$ is:

a.   8kVA
b.   6kVA
c.   10kVA
d.   14kVA.

5   The current in the neutral conductor when all three phases are carrying 20A is:

a.   0A
b.   20A
c.   40A
d.   60A.

Ac motors and dc machines

RECAP

Before you start work on this chapter, complete the exercise below to ensure that you remember what you learned earlier.

1   Draw an impedance triangle for an ac circuit having resistance and inductance connected in series, and identify each side of the triangle.

2   A series circuit consists of a pure resistor of 20Ω and an inductor of negligible resistance. When connected to a 230V, 50Hz supply, the circuit current is 5A. Calculate the voltage across the (a) resistor (b) inductor.

3   For the circuit shown in Figure 8.1 calculate:

a)   the line and phase currents

b)   the power dissipated.

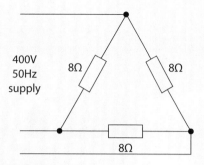

400V
50Hz
supply

8Ω

8Ω

8Ω

Figure 8.1 *Line and phase currents and dissipated power*

4   Draw a phasor diagram, to a scale of 1cm = 15A, for a star connected three-phase balanced circuit with 45A per phase.

LEARNING OBJECTIVES

On completion of this chapter you should be able to:

● Describe the operating principles of dc machines.

● State the basic types and applications of dc machines.

● Explain the operating principles of single-phase and three-phase ac motors.

● Describe the operation of variable frequency drives and inverters.

● Explain the operation of synchronous motors.

● State the basic types, applications and limitations of ac motors.

● Describe the operating principles, limitations and applications of motor control.

Part 1 Basic principle of operation of a dc motor

Rotation

Consider the single-turn coil (which is free to rotate on a shaft) placed inside the fixed magnetic field.

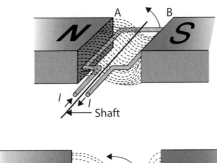

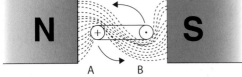

Figure 8.2 *The rotating force produced on a single coil in a magnetic field*

Applying Fleming's left-hand rule, coil side 'A' will have a downwards force produced on it, and coil side 'B' will have an upwards force produced on it. The result of the forces on coil sides A and B is to produce a torque in an anti-clockwise direction and thus produce rotary motion in this direction. The coil will only be forced to a position at 90° to the fixed magnetic field as shown in Figure 8.3.

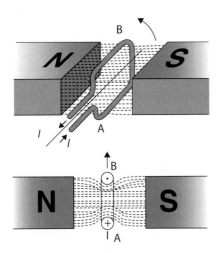

Figure 8.3 *The coil has now rotated 90°*

Commutation

To get past this point a second coil is used and placed at 90° to the first. Current is switched off to the first coil and on to the second. The force on the second coil now rotates the assembly until it is at 90° to the fixed magnetic field. If the direction of rotation is to be maintained the current has to be switched off in the second coil and back on in the first but now in the reverse direction. This switching is carried out automatically with a rotating switch called a commutator.

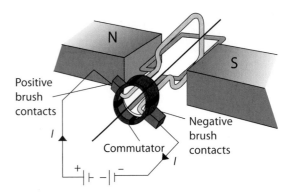

Figure 8.4 *A simple armature arrangement with a four-segment commutator*

The practical dc motor has many armature coils (windings) and a multi-segment commutator.

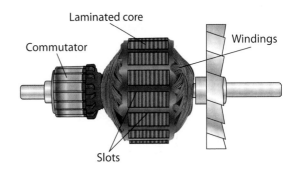

Figure 8.5 *Typical armature*

The commutator segments are insulated from both the shaft and each other.

Try this

Assuming that the coil, Figure 8.2, has rotated 180° from its original position indicate on the diagram below the:

1   coil sides A and B

2   direction of the current induced in each coil side when supplied from the battery in Figure 8.4.

3   main field between the two poles

4   direction of the forces acting on each coil side

5   direction of rotation.

Figure 8.6 *Commutation*

Basic principle of operation of a dc generator

It is important to note that the basic construction of a direct current generator is the same as a direct current motor.

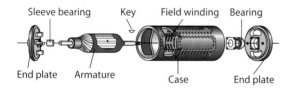

Figure 8.7 *Basic construction of a dc machine*

Both machines are energy convertors. The motor converts electrical energy into mechanical energy whilst the generator converts mechanical energy into electrical energy.

When the single turn coil is rotated within the magnetic field as shown in Figure 8.8 an alternating voltage is produced. This alternating voltage is converted to a direct voltage by the commutator. The commutator rotates with the coil so that the two segments continually interchange with the two carbon brushes, which are stationary. Each end of the coil is connected to a segment of the commutator.

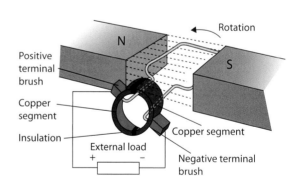

Figure 8.8 *A simple (single-loop) generator*

Generator output

The generated emf in the coil alternates but the output voltage at the brushes retains the same polarity. This is because after 180° rotation the segments interchange their brushes so that as the voltage begins to increase again in the opposite direction the polarity of the output voltage remains the same.

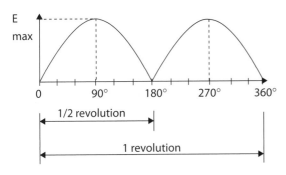

Figure 8.9 *Generator output voltage waveform*

45° some flux is being cut since the coil sides are now moving at an angle to the lines of flux therefore some emf is being generated; the magnitude of the emf depends upon the angle which the coil sides are moving through the lines of flux

90° maximum flux is being cut since the coil sides are moving at 90° (right angles) to the lines of flux therefore maximum emf is generated

135° magnitude of emf generated same as at 45°

180° no emf generated.

![Remember icon] **Remember**

The purpose of the commutator for the motor and generator is:

Motor – to transfer the supply current to the armature coils via brushes and to reverse the direction of the current flowing in the coils as the armature rotates.

Generator – to convert the alternating voltage and current induced in the armature coils into a direct voltage and current at the brushes which are connected to the external circuit.

Magnitude of generated emf

The magnitude of the generated emf depends on the position of the coil within the magnetic field.

Referring to Figure 8.10 we can see that at:

0° no magnetic flux is being cut, since the coil sides are moving parallel to the lines of magnetic flux, therefore no emf is generated

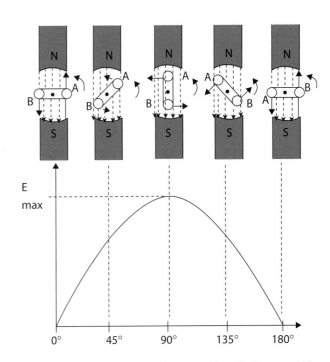

Figure 8.10 *Magnitude of generated emf (during 180° rotation of the coil, and at 45° intervals)*

![Try this icon] **Try this**

A motor converts _____ energy into _____ energy and a generator converts _____ energy into _____ energy.

The commutator is like a _____ switch which _____ the direction of the _____ in the _____ coils of the motor.

The commutator on a dc generator converts the _____ current induced in the _____ coils to a _____ current.

Part 2 Types of dc motor

There are three main types of dc motor: series, shunt and compound wound.

The series motor

The first motor to consider is the series type. As the name implies, the field and armature windings are all connected in series as shown in Figure 8.11.

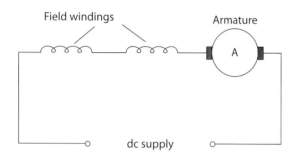

Figure 8.11 *Circuit diagram for a series wound motor*

In a series motor the current flowing in the armature also flows through the field windings, so when the motor is put under load, the armature and field currents increase. This means that the magnetic flux becomes stronger and more torque is produced. When the load is reduced the magnetic flux becomes weaker and the speed increases. In theory, if a series motor was left with no load connected it would continue to increase speed until it destroyed itself.

This type of motor should never be coupled to a load by means of a belt drive, because if the belt breaks the motor's speed will increase rapidly and the motor will disintegrate.

Applications: This type of motor is used where large starting torques are required, such as electric traction cranes and lifts.

To reverse rotation change either the armature or field connections but *not* both.

The shunt motor

This is a motor where the field and armature windings are connected in parallel.

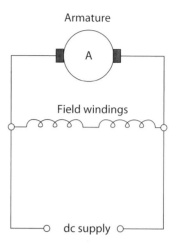

Figure 8.12 *Circuit diagram of a shunt wound motor*

When a motor is connected in this way the field windings receive the full supply voltage across them. As the supply is constant the field strength is constant and therefore the motor speed is fairly constant. As the field strength does not vary, Φ is constant and the speed (n) is proportional to the back emf (E). As the load is increased the armature current and voltage drop also increase producing a slight reduction in motor speed.

Applications: This type of motor is used where virtually constant speed is required on drives, such as small machine tools, fans and conveyor systems.

To reverse rotation change either the armature or field connections but *not* both.

The compound motor

This type of motor has both shunt and series connected field windings, as shown in Figures 8.13 and 8.14.

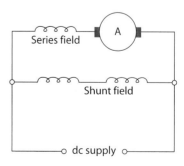

Figure 8.13 *Long shunt connected*

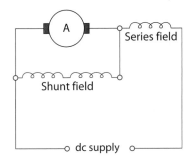

Figure 8.14 *Short shunt connections*

The characteristics of the motor depend on which set of field windings is the strongest. If the series winding is used to increase the magnetic pole strength as the load current increases, this is known as cumulative compounding. This type of motor has characteristics between the series and shunt types, giving high starting torques with safe no-load speeds. This makes cumulative compounded motors ideal for heavy intermittent loads such as lifts, hoists and heavy machine tools.

There are two methods of connecting the shunt windings of a compound motor. Figure 8.13 shows the long shunt connections, whereas Figure 8.14 shows the short connections.

To reverse direction of rotation change either the armature or both sets of field connections but *not* both.

Dc motor starting/speed control methods

One main limitation with starting dc motors is that they draw excessively high starting currents and therefore special type starters have to be used. Older type 'faceplate' starters use resistors to limit the starting current. Modern type electronic starters often use thyristors to limit the starting current.

Older type dc motor speed controllers use variable resistors, modern types use electronic speed controllers.

Try this

When the load on a dc series motor is reduced the magnetic flux becomes _____ and the speed _____ .
A compound motor has a _____ and a _____ field winding.

Types of dc generator

Separately excited generator

As the name implies the field windings have an external dc power supply connected to them to provide the magnetic flux. This supply is often a battery unit which may have a variable resistor incorporated into the circuit to vary the output. The generated output of the machine is taken directly from across the armature.

The output of this type of generator can be fairly constant but the voltage tends to drop off slightly as the load current is increased. This is due to the effect of the load current and armature resistance causing a voltage drop in the armature.

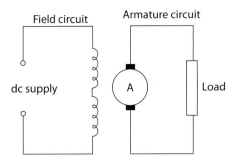

Figure 8.15 *Circuit diagram of separately excited generator*

Self excited generators

Unlike the separately excited generators there is no external supply to power the electromagnets that produce the field. The field windings receive their supply from that produced by the armature. This means that there must be some residual magnetism in the poles for a start so that an initial voltage can be created. Once the armature is turning and current starts to flow in the circuits, the magnetic field becomes stronger and the supply voltage increases. If there is no residual magnetism to start with no voltage can be generated. Similarly if the machine is started so that the residual magnetism is weakened instead of strengthened, it will not excite and therefore not generate.

Connection of the field windings of self excited generators is the same as for dc motors: series, shunt or compound.

Try this

The field windings of a _____ excited dc generator have an _____ dc _____ supply connected to them to _____ the _____.

A _____ excited dc generator will not generate an output _____ if there is no _____ magnetism in the _____ windings.

Part 3 Ac motors

Three-phase ac motors

Three-phase motors are widely used for industrial drives. When they are compared with single-phase motors for similar loads they are less expensive, of smaller dimension, more efficient and generally self-starting.

The current taken by a three-phase motor is between one quarter and one third of that taken by a single phase motor with a similar power rating. This can be seen in Table 8.1. This table shows a (rule of thumb) current per

Figure 8.16 *Three-phase induction motor*

Image courtesy of Brook Crompton UK Ltd

Table 8.1 *Current for ac motors*

h.p.	$\frac{1}{4}$	$\frac{1}{2}$	$\frac{3}{4}$	1	2	3	$5\frac{1}{2}$	$7\frac{1}{2}$	10	15	20	25
kW rating	0.18	0.37	0.55	0.75	1.5	2.2	4.0	5.5	7.5	11.0	15.0	18.5
230V single phase	2.7	4.2	6.3	7.7	12.0	14.4	24.0	32.0	40.0	56.0	76.0	94.0
400V three-phase	0.8	1.2	1.5	2.0	3.6	4.9	8.3	11.0	15.0	21.0	28.0	35.0

phase taken by induction motors (speed 1 440 rev/min) allowing reasonable efficiencies and power factor. When making actual calculations refer to manufacturers' details for precise data.

The power rating of a motor is measured in watts, or more usually, kilowatts. Motor power was formerly measured in horsepower. 1hp = 746 watts or 0.75kW.

Construction of three-phase cage rotor induction motors

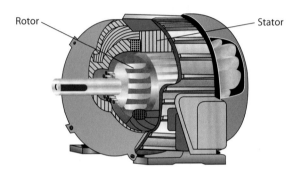

Figure 8.17 *Three-phase cage-rotor induction motor*

This motor has a fixed part, called a stator, which houses the three-phase windings that produce the rotating magnetic field, and a moving part called a cage rotor which revolves within the stator. The stator core is built up from steel laminations with slots to receive the windings. The rotor core is also built up from steel laminations, having longitudinal slots into which lightly insulated copper or aluminium conductors, called rotor bars, are fitted. The rotor bars are short circuited at each end by a heavy copper or aluminium ring so forming a closed circuit.

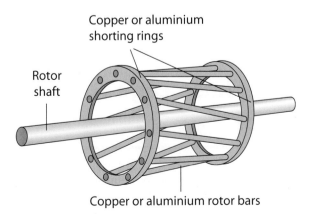

Figure 8.18 *Typical cage rotor construction*

Remember

- The fixed part is the stator and the moving part is the rotor.
- The rotating magnetic field is produced by the stator windings.
- The rotor bars are fitted into slots in the rotor core and are short-circuited by end rings.

Connection of the three-phase cage-rotor induction motor

Let's take a look inside the terminal box of this three-phase cage rotor motor. There are six stud type terminals moulded into an insulated terminal board to which each end of the three-phase windings are connected as shown below.

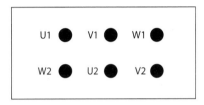

Figure 8.19 *Terminal board*

This enables the windings to be connected either in star or delta configuration.

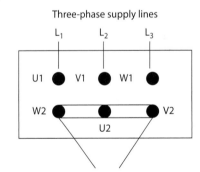

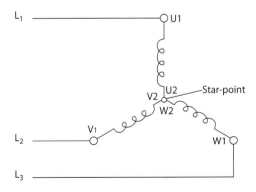

Figure 8.20 *Star connection*

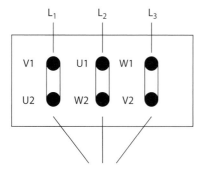

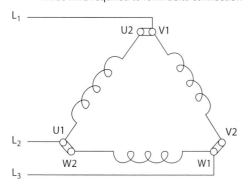

Three links required to form delta connection

Figure 8.21 *Delta connection*

Many three-phase cage rotor motors have just three terminations inside the terminal box. The windings in this case will be permanently connected in either star or delta configuration. It is not normally possible to change the connections as they are internally connected.

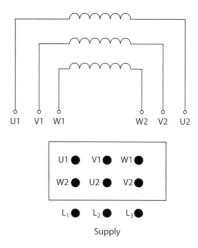

Figure 8.22 *Windings in delta connection*

Try this

Draw in the connections so that the windings in Figure 8.22 are connected in a delta configuration and indicate where the main supply would be connected.

Part 4 Operation of the three-phase cage rotor induction motor

When the three-phase ac supply is connected to the stator windings a rotating magnetic field is set up by the windings.

This rotating field cuts through the rotor bars and induces an emf in them. Since the rotor bars are short-circuited by end rings, considerable induced currents flow in them

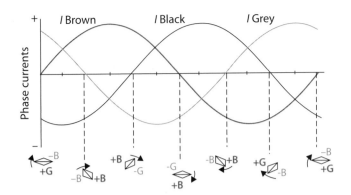

Figure 8.23 *Rotating magnetic field from a three-phase supply*

which in turn create their own magnetic fields. These fields interact with the stator's rotating magnetic field and exert magnetic forces upon the rotor bars. These forces cause the rotor to turn in the same direction as that of the rotating magnetic field.

The rotor never quite manages to catch up with the synchronous speed of the rotating magnetic field, if it did, then no emf and current would be induced in the rotor bars (since there must be a relative motion between the rotating field and the rotor bars to create a torque) and the rotor would tend to slow down and stop.

The rotor will settle down to a steady speed just below synchronous speed. This difference between synchronous speed and the actual rotor speed is called the slip speed, since the rotor tends to slip behind the rotating magnetic field of the stator.

Reversal of rotation

The standard sequence of a three-phase supply is brown, black and grey or L_1, L_2 and L_3.

The direction in which the magnetic field rotates is directly related to the sequence that the phases are connected to the stator windings. Rotation of the field can be reversed by reversing the connection of any two incoming phases. As a result the rotation of the rotor will also be reversed.

Advantages of a cage rotor induction motor

- Due to the simple construction (only the stator winding is connected to the supply) the motor is comparatively cheap.
- Due to the cage construction of the rotor they are mechanically strong and robust and therefore they are particularly useful for industrial drives.
- They are normally self-starting (i.e. no special starting equipment is required).
- They require little maintenance as there are no rubbing contacts (brushes) on the rotor.
- Good speed/load characteristic.

Limitations

- High starting current (up to 6 or 10 times full load current).
- Due to the inductance of the stator winding it has a low power factor (typical 0.7). This is even worse at the moment of starting.
- Poor starting torque and therefore can only be started with light loading.

Typical applications: Machine tools, industrial drives and small pumps.

Try this

To change direction of _____ change any _____ phases.

The _____ magnetic field is set up by the _____ windings.

The _____ fields (stator and _____) cause the rotor to turn.

The _____ turns in the _____ direction as the rotating _____.

Starting torque of cage rotor motors

The starting torque depends on the design of the cage rotor.

One main drawback with the cage rotor motor is that it has a relatively high starting current and a low starting torque.

A double cage rotor motor has a lot better starting torque than a single cage rotor motor.

In Figure 8.24:

(A) with single cage rotor
(B) with double cage rotor.

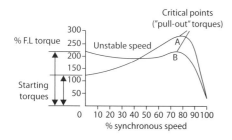

Figure 8.24 *Typical torque/speed characteristics curves for three phase induction motors*

In each case the knee of the curve indicates the point where the motor pulls-out owing to the torque increasing above a certain value. If the motor is loaded beyond this point it will no longer take the load and the speed will fall quickly to zero.

Remember

The single cage rotor has a low starting torque and a high starting current.

Choose a double cage rotor for a better starting torque.

The double-cage rotor motor

The double-cage rotor motor is designed to provide a high starting torque with a low starting current. The rotor is so designed that the motor operates with the advantage of a high resistance rotor circuit during starting (outer bars), and a low resistance rotor circuit under running conditions (inner bars).

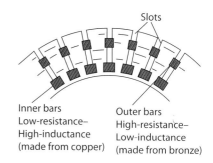

Figure 8.25 *Section of a double-cage rotor*

Typical applications: Conveyors and industrial drives.

Three-phase wound rotor induction motor

The stator construction is identical to that of the cage rotor motor, as previously described.

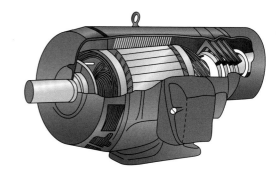

Figure 8.26 *Three-phase wound rotor induction motor*

The rotor consists of three windings having many turns connected in star or delta configuration, the terminations of which are brought out and connected to three slip rings fitted on the rotor shaft.

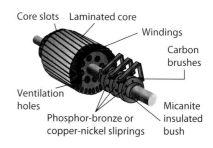

Figure 8.27 *Wound rotor with slip rings*

The slip rings allow connection of the rotor windings to external resistances as shown on Figure 8.28.

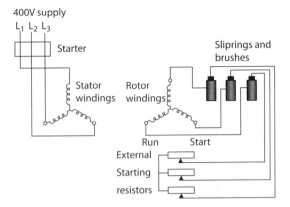

Figure 8.28 *Wound rotor motor connections*

Operation

The motor is started with all the external resistance connected into the rotor circuit. Then the resistance is gradually reduced as the motor gains speed and at full speed is cut out altogether. The slip rings are then shorted out and the motor continues to run as a cage motor.

In large motors, which sometimes run almost continuously, a brush lifting device is incorporated with the shorting gear to eliminate needless brush and slip ring wear. The wound rotor, like the cage rotor has no external electrical connections to the supply. For some applications the rotor slip rings are not shorted out and the rotor resistance is used as a form of speed control.

Typical applications: Large pumps and compressors.

Changing the direction of rotation

The direction of rotation may be changed by changing any two phases.

Synchronous motors

A synchronous motor is basically an ac motor that has a magnetic rotor.

Three-phase synchronous motor

The stator windings of the three-phase synchronous motor are exactly the same as in the induction motor, so that when it is connected to the three-phase supply, a rotating magnetic field is produced. Instead of having a cage rotor it has a rotor with a dc excited winding, which is designed to cause the rotor to lock-on or synchronize with the rotating magnetic field produced by the stator windings. Once the rotor is synchronized it will run at the same speed as the rotating magnetic field despite any variation in load.

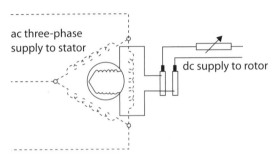

Figure 8.29 *Three-phase synchronous motor with dc supply to the rotor*

One main limitation with this type of motor is that it is not self-starting and methods have to be employed to get the motor started. The method used to start both the salient-pole and synchronous induction types is to start them as an induction motor.

Synchronous motors are often installed not only to drive some constant load, such as a pump or fan, but also for power factor correction. The power factor of the motor can be adjusted by varying the rotor excitation. Unity power factor operation is possible and by using a high rotor excitation the motor can operate at a leading power factor. Therefore a large synchronous motor can be used to compensate for the lagging power factors of other inductive loads on the same plant.

A small single-phase synchronous motor (as used for electric clocks) has a permanent magnet rotor.

Try this

When both ends of each _____ of a three-phase motor are brought out to the _____ box the motor can be connected in _____ or _____ .

The three-phase _____ rotor induction motor has a _____ starting current and a _____ starting torque.

The _____ cage rotor motor has a _____ starting current and a _____ starting torque than a _____ cage rotor motor.

The _____ rotor motor is started with _____ the _____ resistances in the rotor circuit, and at full speed they are _____ .

The _____ motor is not _____ starting but it can be used for _____ correction.

Part 5

Slip

We have already established that it is not possible for a normal induction motor to run at synchronous speed.

If torque is to be produced the induced current is necessary, therefore the rotor must run at something less than the synchronous speed of the rotating field.

The difference between the actual rotor speed (N_r) and synchronous speed (N_s) is called the slip speed.

It is more usual to speak of slip of an induction motor as a fractional (per unit) value which is given the symbol s.

$$S = \frac{N_s - N_r}{N_s} \text{ or as a percentage value}$$

$$S(\%) = \frac{N_s - N_r}{N_s} \times 100.$$

Remember

There must be relative movement between rotor conductors and the rotating magnetic field for torque to be produced.

Speed, poles and frequency (three-phase winding)

A three-phase stator winding consists of three sets of coils evenly distributed around the core of the stator and connected in either star (Y) or delta (Δ). Each of these windings could have two or more poles per phase depending upon the speed required.

The rotating magnetic field set up by a two-pole winding completes one revolution (360°) in one complete cycle of the mains supply, whilst with a four-pole winding (2 pairs of poles) it completes one revolution in two cycles.

So, as the number of poles per winding is increased so the speed of the rotating magnetic field within the machine decreases (see Table 8.2).

Table 8.2 *Synchronous speeds and standard rotor speeds at full load (50Hz supply)*

Poles	Synchronous speeds revs/min	Rotor speeds revs/min
2	3 000	2 900
4	1 500	1 440
6	1 000	960
8	750	720
10	600	580
12	500	480
16	375	360

Synchronous speed

The synchronous speed of ac motors depends on the frequency of the supply and the number of pairs of poles on the stator.

Synchronous speed can be calculated by using:

$$N_s = \frac{f}{P} \times 60.$$

- N_s is the synchronous speed in rev/min
- f is the frequency in Hertz (cycles per second)
- p is the number of pairs of poles

(60 converts frequency to cycles per minute so that the speed can be calculated in rev/min).

Or use:

$$n_s = \frac{f}{P}$$

Where n_s is the synchronous speed in revolutions per second (rev/s).

Speed and slip calculations

Example:

We need to calculate the synchronous speed in rev/min of a 16 pole motor if the supply frequency is 50Hz.

$$N_s = \frac{f}{P} \times 60 = \frac{50}{8} \times 60 = 375 \text{ rev/min}$$

Remember

p = number of pairs of poles in these calculations.

Try this

Determine the synchronous speed in rev/s of a 4-pole motor connected to a 50Hz supply.

Example:

A twelve-pole, 50Hz induction motor runs at 475 rev/min. Calculate:

a. the synchronous speed

b. the percentage slip.

a. $N_s = \frac{f}{P} \times 60 = \frac{50}{6} \times 60$

$= 500 \text{ rev/min}$

b. $S(\%) \frac{N_s - N_r}{N_s} \times 100 = \frac{500 - 475}{500} \times 100$

$= \frac{25}{500} \times 100$

$= 0.05 \times 100$

$= 5\%$

Example:

An induction motor has six poles and a per unit slip of 0.05 at full load when the supply frequency is 50Hz. To determine the rotor speed of the motor we must first calculate the synchronous speed.

$$n_s = \frac{f}{P} = \frac{50}{3} = 16.67 \text{ rev/s}$$

Now using the transposition of $S = \frac{n_s - n_r}{n_s}$ so that:

$$n_r = n_s(1 - s)$$

$$= 16.67(1 - 0.05)$$

$$= 16.67 \times 0.95$$

$$= 15.84 \text{ rev/s}$$

Try this

An eight-pole induction motor runs at 12 rev/s when supplied at 400V, 50Hz. Calculate the percentage slip.

_____

_____

Try this

A two-pole induction motor has a slip of 0.03 per unit when operating at full load on a 50Hz supply. Calculate the:

1   synchronous speed

_____

_____

2   rotor speed

_____

_____

Slip speed is the difference between the rotor speed and the synchronous speed.

The _____ speed of ac motors depends on the _____ of the supply and the _____ of _____ of _____ .

Part 6 Operation of single-phase ac induction motors

Figure 8.30 _Single-phase induction motor with exposed cage rotor and stator windings_

As we have seen, a three-phase induction motor is self-starting due to the rotating magnetic field of the stator. This magnetic field cuts the rotor conductors and exerts a force on them which causes the rotor to turn.

However, a simple single-phase induction motor with a single winding is not self-starting. If the single winding is supplied with ac it simply produces a pulsating field which rises and falls with the alternating current and so this type of field produces no torque in the rotor.

Surprising as it may seem we can actually start this simple motor by spinning the rotor. The rotor conductors are still being cut by the pulsating field, but in addition, the conductors are moving through the field. We then have two fields (rotor and stator) interacting with each other and the same conditions exist as in a rotating field. Once started the rotor will continue to run in the same direction as we turned the rotor to start it.

Obviously this arrangement is most unsatisfactory and a means of making single-phase motors self-starting has to be considered. The basic principle is to get the stator field to move (rotate) so that the rotor may follow. Several methods are employed and all of them use an additional winding. This additional winding is called the start winding and it is fitted to the stator of the motor.

The start winding is usually short-time rated and would overheat if left in the circuit for more than a few seconds.

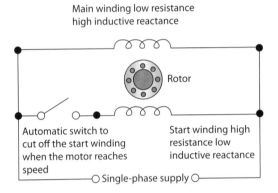

Main winding low resistance
high inductive reactance

Rotor

Automatic switch to cut off the start winding when the motor reaches speed

Start winding high resistance low inductive reactance

Single-phase supply

Figure 8.31 *The auto-switch is normally a centrifugal switch*

The magnetic fields produced by the out of phase currents create the necessary torque on the rotor to make the motor self-starting when connected to the single phase supply.

Two motors operating on this principle are the split-phase and the capacitor split-phase motor. The approximate phase displacement between the currents is 30° for the split-phase and 90° for the **capacitor start** split-phase.

Types of single-phase induction motor

Split-phase motor

The split-phase motor has two stator windings and they are:

- a main winding which is termed the run winding
- an auxiliary winding called the start winding.

Windings connections

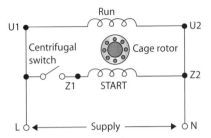

Figure 8.32 *Split-phase motor connections*

The run winding is wound with heavier gauge wire than the start winding and has less resistance than the start winding, which is wound with very fine gauge wire. The inductive reactance of the run winding is however greater than that of the start winding.

As the inductive reactance of the run winding is greater than that of the start winding the current in the two windings will be out of phase with each other and reach maximum values at different times during each cycle. The magnetic flux produced by these currents will also reach maximum and minimum values at different instants of each cycle. The start winding is short-time rated and will burn out if left in circuit. To prevent this happening it is automatically cut out by a centrifugal switch when the motor reaches about 75% full speed.

The moving arms for the centrifugal switch are fitted on the cage rotor shaft (Figure 8.33) and operate contacts fitted on the inside of the stator frame.

Centrifugal switch mechanism

Figure 8.33 *Cage rotor and centrifugal switch*

Changing the direction of rotation

To change direction of rotation, reverse the connection of one winding only, i.e. start or run.

Typical starting torque – 175–200% of full load torque

Starting current – approximately 600–900% of full load current.

Applications: Domestic appliances, such as washing machine agitators, tumble dryers.

Capacitor-start, induction-run motor

The starting characteristics of a split-phase motor can be improved by connecting a capacitor in series with the start winding. This type of motor is called a capacitor start-induction run motor.

Windings connections

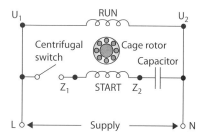

Figure 8.34 *Capacitor-start, induction-run motor*

The phase displacement in the currents of the two windings at starting is now approximately 90°; therefore, the magnetic flux set up by the two windings is much greater at starting than in the split-phase type motor, and this produces a relatively higher starting torque. The centrifugal switch operates the same as with the split-phase motor but now it disconnects the start winding and its series connected capacitor.

Application: Compressors, refrigerators and other applications involving a hard starting load.

Changing direction of rotation: reverse start or run winding connections – *not* both.

Universal (ac/dc) motor

The universal motor can be operated on either an alternating or direct current supply and is very similar in construction to the dc series wound motor.

The armature and the field windings are connected in series exactly like the dc series motor as shown in Figure 8.35.

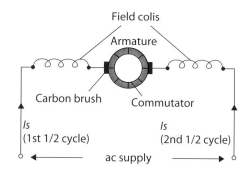

Figure 8.35 *Universal motor*

Operation

The basic principle of operation of the universal motor is the interaction between the main field and the armature field, which is produced by the same current (the supply current I_s).

First half cycle

I_S flows through the left-hand side field coil, then it passes into the armature conductors via the commutator and then through the right hand side coil.

Second half cycle

I_S flows in the opposite direction. Because the supply current is alternating, both the main field polarity and armature conductor polarity will change at the same moment in time, therefore the motor will continue to run in the same direction.

Figure 8.36 *Universal motor armature*

Advantages

- Can be used on ac or dc.
- Higher running speeds than the induction type motor (typical 8 000 to 12 000 rev/min maximum).

- Cheap to produce.
- High starting torque.
- The higher running speed plus a good power factor develops more output power for a given size than any other single phase motor.

Limitations

- Commutation on ac is not as good as on dc.
- Brush and commutator wear is more rapid when used on ac due to additional sparking at the brushes caused by the effect of inductance.
- Brushes are also prone to wear, especially when the motor is running at high speeds for long periods.

- If the load on the motor is reduced the speed will rise rapidly.

Application: Universal motors are widely used in domestic appliances such as vacuum cleaners, food mixers, portable tools such as drills and saws.

These motors do not usually exceed 1/4 or 1/3 horsepower.

Changing the direction of rotation

To reverse direction of rotation change either the field coil connections or armature connections – *not* both.

Try this

The split-phase induction motor has _____ windings, a _____ winding and a _____ winding connected in _____ across the supply.

On a split-phase motor the _____ cuts out the start winding to prevent it from _____ out.

To change the direction of rotation of a capacitor-start induction-run motor reverse _____ the _____ or _____ winding connections.

_____ motors can operate on either ac or dc supplies, and can run at very _____ speeds.

Part 7 Control of ac motors

Ac motor starters

Motor starters must all be able to:

- connect and disconnect the motor to the supply in a safe manner
- give the motor protection from sustained overload.

Motor starters should be able to:

- prevent the motor from restarting after a supply failure or severe undervoltage (essential except for a few special applications)

- provide means of current limitation for starting larger motors.

Where applicable motor starters should also:

- control the starting torque
- reverse the direction of rotation of the motor
- control the speed of the motor.

Types of three-phase motor starter

There are several types of starter used for starting three-phase cage rotor induction motors including: direct-on-line, star-delta, rotor-resistance, soft-start and variable frequency.

Direct-on-line (DOL) starter

Direct-on-line starters are used for starting the majority of small three-phase cage rotor induction motors.

With this type of starter the stator windings of the motor are connected directly to the supply lines L_1, L_2 and L_3. Therefore there are 400 volts across each stator winding on starting. Since the motor is at rest when the supply is switched on, the initial starting current is heavy and may cause some disturbance to the electricity supply (for example lights could flicker or dim). To overcome these problems supply companies limit the use of DOL starting of induction motors above certain power ratings, typically 7.5kW. Where doubt exists a check should be carried out to ensure direct-on-line starting is permissible.

The initial current surge on starting can be six to ten times the full load current and the initial starting torque is about 150% of full load torque.

Application: Direct-on-line starters are suitable for light starting loads.

Remember

All motor starters consist of a mains circuit, which switches the windings of a motor, and an auxiliary circuit which is used to control the switching functions.

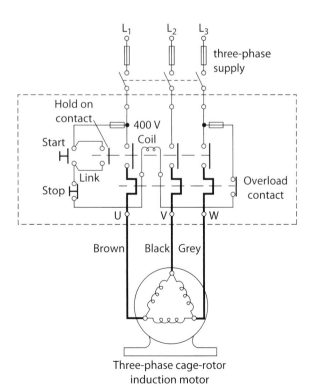

Three-phase cage-rotor induction motor

Figure 8.37 *Direct-on-line starter*

In Figure 8.37 the main circuit is shown by thick lines, the control (auxiliary) circuits shown by thinner lines, the fuses provide short-circuit protection and the overloads may be thermally or magnetically operated.

Operation of direct-on-line starter

Press the start button and the contactor coil energizes. The main contacts and hold-on contact close and the motor starts to run.

When the start button is released the contactor coil remains energized via the hold-on contact and the motor continues running.

Press the stop button the contactor coil de-energizes, the main contacts open and the motor stops.

Overload protection

If the motor draws excessive current due to overload the overload coils will trip out the overload contact, which opens the circuit to the contactor coil. The coil de-energizes, the main contacts open and the motor stops. Should one phase fail the motor would be put under undue pressure to continue to work. This situation is called single phasing and the overload equipment should detect this and stop the motor.

All motors having a rating exceeding 0.37kW must have control equipment with overcurrent protection.

No-volt protection

If there is a supply failure, the contactor coil will de-energize and the hold-on contact will open. Hence, the contactor coil cannot re-energize again (when the supply is reinstated) until the start button is pressed by the operator.

Undervoltage protection

If the supply voltage falls to about 80% of its nominal value the contactor coil de-energizes and trips out the contactor. The operator will have to press the start button to re-energize the contactor because the hold-on contact will be open again.

It is most important that no-volt/undervoltage protection is provided to prevent automatic starting of machinery after a failure of the supply. For example a machinist could be in serious danger if the lathe restarted unexpectedly when the supply was re-instated.

Remote start/stop control of a direct-on-line starter

When connections are required for remote start/stop buttons these extra start buttons are wired in parallel with existing start button and the extra stop buttons are wired in series with existing stop button.

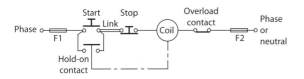

Figure 8.38 *Control circuit for direct-on-line starter*

When the control circuit is single phase (connected between the line and neutral) the fuse F2 can be omitted or replaced by a link.

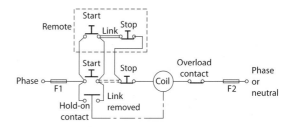

Figure 8.39 *Control circuit with remote start/stop*

Only three wires are required to connect in the remote start/stop control. In order to add on another remote start/stop position simply remove the link, run three more wires out and connect as above.

Forward/reverse DOL starters have a forward and a reverse contactor which are interlocked mechanically and electrically to prevent them both being energized at the same time.

Try this

Direct-on-line starters are suitable for _____ starting loads.

A DOL starter has three _____ , one in each line, to protect the motor when it starts drawing _____ current.

_____ protection is provided to prevent _____ of machinery after a supply _____ .

When connecting _____ stop/start controls the start buttons are connected in _____ and stop buttons are connected in _____ .

Part 8 Reduced voltage starting methods

Star-delta starting

This is one of the most common methods of reduced voltage starting. On starting, the stator windings are first connected in a star configuration. This is to reduce the voltage across each winding.

i.e.

$$V \text{ phase} = \frac{V_L}{\sqrt{3}} = \frac{400}{1.732}$$

$$= 230V \text{ approximately}$$

230V is approximately 58% of 400V.

The motor is then run up to speed with 230V across each winding and when it has attained approximately full speed the starter is switched to connect the windings in delta configuration with the full line voltage of 400V across each winding.

Sufficient time must be allowed for the motor to run up to speed before switching from star to delta to prevent the possibility of heavy overloads and damage to the motor when using this method.

Typical values for starting a three-phase cage rotor induction motor by star delta are:

● initial starting current 2 to 4 times full-load-current

● initial starting torque 50% of full-load-torque.

The starting current and starting torque is reduced to one third of that which would occur if the motor was started direct-on-line.

Application: Used for starting on no-load or light load. Motors of all sizes up to about 13kW can be star delta started.

Manual star delta starter

This method of star delta starting requires the operator to manually switch the starter into a star position and then when the motor runs up to speed change over to delta. Older manual type starters are now being replaced by the newer automatic types.

Auto star delta starter

Unlike the manually operated starter the auto star delta starter only requires a push of the start button to set the process in motion.

Operation

Press the start button and the main contactor coil energizes.

- The star contactor coil and timer (T) both energize via the timer contact (Y) and delta interlock contact.
- The main contacts close and connect the three-phase supply lines to the motor winding terminals U_1, V_1 and W_1.
- The star contacts close and connect motor winding terminals, U_2, V_2 and W_2, together to form the star point.
- The motor starts to run on reduced voltage with 230V across each winding.

Release the start button and the main contactor coil remains energized via the hold-on contact.

- The star contactor remains energized via the hold-on contact and the timer contact (Y). The motor continues to run on reduced voltage.
- The timer times-out and switches the change-over contact from the Y position to the Δ position.
- The star contactor coil de-energizes and the delta contactor coil energizes via the timer contact D and the star interlock contact.

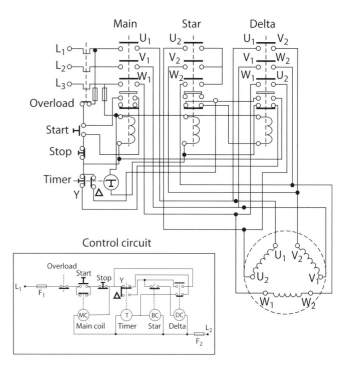

Figure 8.40 *Auto star delta starter*

- The delta contacts close and connect the following motor winding terminals together, U_1 to V_2, V_1 to W_2 and W_1 to U_2, to form the delta connection of the windings.
- The motor continues to run on full supply voltage with 400V across each winding.
- The timer de-energizes (resets) to the star position, and the delta contactor coil remains energized via its own hold-on contact.

Press the stop button and the main and delta contactor coils de-energize, the main contacts open and the motor stops.

Overload protection

If the motor draws excessive current the overload coils will trip out the overload contact to de-energize the contactor coils.

No-volt and undervoltage protection

If either the supply fails or the supply voltage falls drastically the contactor coils will de-energize and they cannot re-energize until the operator has pressed the start button because the hold-on contact opens the circuit to the coils when the main contactor de-energizes.

Try this

A three-phase cage-rotor induction motor is often started using a star delta starter to reduce the _____ , however, the _____ torque is also reduced. On starting, the _____ windings are connected in _____ and when the motor has reached approximately _____ the _____ windings are connected in _____ .

Part 9

Starting three-phase motors on load

The starters previously mentioned were used to start three-phase cage rotor induction motors with no-load or light loads. When it is required to start a motor against a heavy load, a three-phase slip-ring motor with a wound rotor can be used. This is started by a rotor resistance type starter as shown in Figure 8.41.

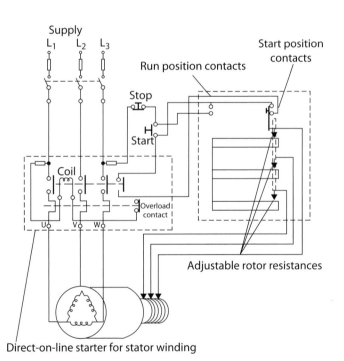

Figure 8.41 *Hand operated rotor resistance starter*

Wound rotor motor starters

The starting torque of this type of motor depends upon the total resistance in the rotor circuit, therefore the external resistances can be arranged to give maximum motor torque on starting from standstill.

However, the starting torque is usually kept down to about 150% of full-load-torque, with a starting current of about 150% of full-load-current. This is done to reduce the disturbance to the mains supply and the shock to the driven machine.

Remember

Rotor resistance starters are used for starting three phase slip-ring motors with wound rotors.

Solid-state 'soft start' controller/starters

These controllers are designed so that they not only limit the inrush current when the motor is first switched on, but also control smooth acceleration and torque build up to meet the requirements of the load.

There are several variations on types of solid state starters but all have similarities. One of the most common types comprises three pairs of back-to-back thyristors connected in series with the three-phase supply lines, as shown in Figure 8.42 below. (Thyristors are covered in more detail in Chapter 12.)

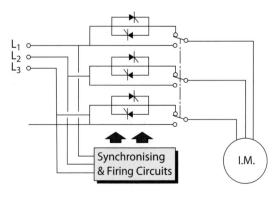

Figure 8.42 *Typical thyristor-controlled soft-starter*

Operating principle

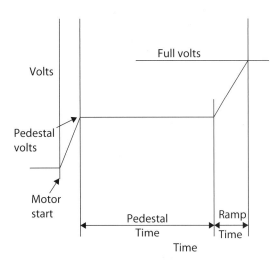

Figure 8.43 *Soft-start operating principle*

When the motor is first switched on the voltage applied is increased to the pedestal voltage and it is maintained at this level for what is known as the pedestal period of time. This is so that the starting current is controlled whilst the voltage is enough to allow the motor to generate the necessary breakaway torque.

Then the controller automatically moves on to the ramp stage. The voltage now increases allowing the motor to accelerate smoothly up to full speed in the running condition with the full voltage supplied to it. By automatically ramping down the voltage the motor can also be stopped softly.

Single-phase direct-on-line starter

The operation of this starter is the same as for the three-phase direct-on-line starter in Figure 8.37, since the control circuit is the same. The operating coil is 230V and requires a neutral connection.

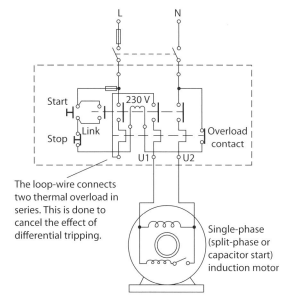

Figure 8.44 *Single-phase DOL starter*

The loop-wire connects two thermal overload in series. This is done to cancel the effect of differential tripping.

Single-phase (split-phase or capacitor start) induction motor

---

Try this

One of the most common methods of _____ voltage starting is _____ starting.

With this method, the _____ windings are connected in _____ to reduce the _____ across each _____ on starting, and then when the motor has attained approximately _____ speed the starter is switched to connect the windings in _____ with the full _____ voltage of _____ across each _____.

Rotor _____ starters are used for starting _____ rotor motors against _____.

Soft starting limits the inrush _____ when the motor is _____ switched on and also gives _____ acceleration and _____ build up to meet the requirements of the _____.

Part 10

Speed control of ac induction motors

There are four main factors which affect the speed of induction motors:

- The number of poles on the motor.
- The supply frequency.
- The supply voltage.
- The load on the motor.

Speed control may be achieved by varying the supply voltage if the frequency and the number of poles are fixed. However, this will also affect the torque produced by the motor, since torque is proportional to the voltage squared ($T \propto V^2$), which in turn affects the amount of load the motor can drive.

Alternatively, if the supply voltage is fixed, speed control can be achieved by varying the frequency with a control system which incorporates an inverter.

Inverters and variable-frequency drives

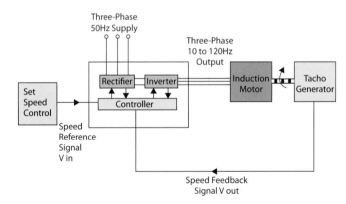

Figure 8.45 *Simple block diagram of a closed-loop inverter-fed variable-frequency drive induction motor speed control system*

Operating principle

- The two signals (speed feedback and speed reference) are compared inside the controller and any difference between them produces an error signal which is used to control the system.

- If V_{out} is greater than V_{in} the speed is reduced by reducing the inverter output frequency and if V_{out} is less than V_{in} the speeds is increased by increasing the inverter output frequency.

- One main advantage with this system is that induction motors can run at speeds in excess of the normal mains frequency operating speed.

- The rectifier changes the ac supply to dc and the inverter produces a variable frequency ac output from the dc input.

Inverter supplies

Many inverters require a three-phase supply and provide a three-phase output but some only require a single phase supply and also provide a three-phase output. These are used with smaller sized three-phase motors up to about 5kW.

Very small inverters are available for use with single-phase motors, usually less than 1kW.

Variable-frequency inverter-fed induction motors are now used in ratings up to hundreds of kilowatts.

Remember

Rectifiers convert ac to dc whereas inverters convert dc to ac.

Variable frequency drive (VFD) motors

The motor used in a VFD system is usually a three-phase induction motor; however, some types of single phase and synchronous motors can be used. Motors that are designed for fixed-speed operation are often used because they offer higher reliability and better VFD performance.

Advantages of a (inverter controlled) VFD

- Continuous smooth operation of motors.
- Accurate speed control.

- Speeds well above synchronous speed (3 000 rpm) attainable.
- Three phase motors can be run off a single-phase supply.
- Solid state switching.
- High efficiencies.
- Regenerative braking (energy efficient).
- Typical starting torque is 150% of the motor's rated torque.
- Typical starting current is less than 50% of the motor's rated current.

Limitations of (inverter controlled) VFD

- Harmonics (distorted waveforms may be transmitted to the mains supply).
- Cogging (the motor jerks and vibrates) at low motor speeds.
- Motor overheating if continuously run at slow speeds.

Applications of (inverter controlled) VFD

- Process control in manufacturing.
- Ventilation system fans (large buildings).
- Machine tool drives.
- Conveyors.
- Compressors.
- Pumps.

Congratulations you have now completed Chapter 8. Correctly complete the self-assessment questions before you progress to Chapter 9.

SELF ASSESSMENT

Circle the correct answers.

1   A dc machine which has two different field windings is:

   a.   series wound

   b.   compound wound

   c.   shunt wound

   d.   separately excited.

2   A single-cage-rotor motor has a:

   a.   high starting current and a high starting torque

   b.   low starting current and a low starting torque

   c.   high starting current and a low starting torque

   d.   low starting current and a high starting torque.

3   When starting a motor direct-on-line the stator windings are first connected in star and the voltage across each winding will be reduced to approximately:

   a.   50%

   b.   58%

   c.   75%

   d.   80%.

4   The synchronous speed of a six-pole induction motor operating from a 50Hz supply is:

   a.   500 rev/min

   b.   1 500 rev/min

   c.   1 000 rev/min

   d.   600 rev/min.

5   Which of the following is *not* an advantage for an inverter motor controller?

   a.   smooth motor operation

   b.   solid state switching

   c.   high running speeds

   d.   poor speed control.

Electrical components

9

RECAP

Before you start work on this chapter, complete the exercise below to ensure that you remember what you learned earlier.

1 Describe how the direction of rotation can be reversed on a shunt wound dc motor.

_____

2 State the function of the sliprings on a wound rotor induction motor.

_____

3 How are extra start and stop buttons connected to a DOL starter?

_____

4 Calculate the synchronous speed of a twelve-pole ac motor when connected to a 50Hz supply.

_____

LEARNING OBJECTIVES

On completion of this chapter you should be able to:

● Specify the main types of the stated electrical components.

● Explain the operating principles of these electrical components.

● Describe how these electrical components are applied to electrical systems and equipment.

● State the advantages and disadvantages/limitations of these electrical components.

In this chapter we will be looking at relays, contactors, solenoids, over-current protection devices, RCDs and RCBOs.

Part 1 Relays

A relay is an electrically operated switch that opens or closes contacts to control the flow of current in one or more separate circuits.

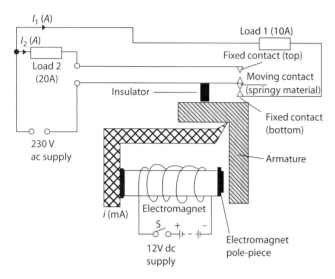

Figure 9.1 *Simple electromagnetic relay*

Operation of a relay

When switch 'S' is closed, current 'I' flows in the relay coil and the magnetic field set up by this current attracts the armature to the electromagnet's pole-piece. The armature forces the moving contact 'away' from the 'bottom' fixed contact and 'up' to the 'top' fixed contact. In doing so, current 'I_1' ceases to flow through load 1 and current 'I_2' flows through load 2. (The relay's function in this circuit is that of a 'changeover' switch.)

Reasons for using a relay

There are many reasons for using relays which include:

- Control circuits, carrying only a few milliamps, may be used to switch larger currents.
- The control circuit may have a different type of supply to the main circuit.
- An extra low voltage control circuit can be used to switch a low voltage circuit.
- For safety reasons a control circuit can be 'electrically isolated' from the main circuit.

A practical relay will have many normally open (**N/O**) and normally closed (**N/C**) contacts.

Image courtesy of Omron Electronics Ltd.

Figure 9.2 *General purpose plug-in relay*

Common relay contact configurations

SPST – single-pole single throw
SPDT – single-pole double throw
DPST – double-pole single throw
DPDT – double-pole double throw
B-M means 'break before make'.

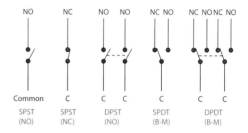

Figure 9.3 *Relay contact configurations*

Solid state relay

A solid state relay (SSR) is purely an electronic switching device which normally uses a small control signal to switch loads carrying larger currents or operating at other voltages.

It has no moving parts with the contacts being replaced by transistors, thyristors or triacs, hence it will switch loads on and off a lot faster than electromechanical type relays.

Finder SpA

Figure 9.4 *Solid state relay*

Contactors

Contactors are electrically operated switches which are used for switching (high current) power circuits. They are similar to a relay; however, relays are normally used to switch circuits carrying lower currents.

Basic operating principle of a contactor (similar to a relay)

When a small current passes through an electromagnet a magnetic field is produced, this attracts an armature and closes the moving contacts against the fixed contacts. The force developed by the electromagnet holds the moving and fixed contacts together.

When the contactor coil is de-energized, the magnetic field is broken and springs force the two sets of contacts apart.

Applications: Contactors are widely used to control electric motors, heating loads, lighting loads, capacitor banks and other electrical loads (especially on three-phase systems).

Figure 9.5 *A typical contactor*

Choice of relay or contactor

When choosing a relay or contactor for a particular application it is important to ensure that:

- the rated operating voltage and current (ac or dc) are suitable for the type of load they are switching
- the voltage rating (ac or dc) of the operating coil is suitable
- the contact configurations suit the circuit requirements.

Solenoids

A coil, consisting of a number of turns of wire wound in the same direction, which is capable of carrying a current is called a solenoid.

Figure 9.7 *Basic electromagnet*

Electromagnet

When the coil of a solenoid is wound around an iron or steel core and a current is passed through it, the coil will cause the core to act as a magnet. This is called an electromagnet or an electromagnetic solenoid.

This type of solenoid is used for operating relays, contactors, valves and trembler bells, etc.

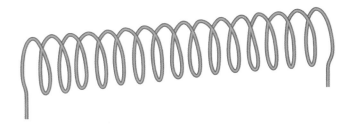

Figure 9.6 *Basic solenoid*

Figure 9.8 *Electromechanical solenoid*

Electromechanical solenoid

An electromechanical solenoid usually has a coil wound around a movable iron or steel plunger.

The magnetic overload trip in a three-phase direct-on-line motor starter has three of these types of solenoid, one for each phase.

Try this

The relay is an electrically _____ device that _____ and/or closes _____

to control the _____ in a separate _____. When current flows through the relay _____

it sets up a _____ field which attracts the _____. When the _____

moves it can open and/or _____ contacts in the main _____ .

A contactor is an _____ operated switch normally used for switching higher _____ circuits.

A coil of _____ which is capable of carrying _____ is called a _____ .

Overcurrent protection devices

To comply with BS 7671, overcurrent protection devices have to be installed in an electrical installation to protect:

- persons and livestock against injury
- property against damage due to excessive temperatures or electromechanical stresses caused by any overcurrents likely to arise in live conductors.

The terminology and characteristics of protective devices

Fusing current: This is the minimum current causing the fuse to blow.

Current rating: This is the maximum continuous current-carrying capability of a fuse, without it deteriorating.

Fusing factor: The level of protection offered by the protective device is given by the fusing factor.

$$\text{Fusing factor} = \frac{\text{fusing current}}{\text{current rating}}$$

Example:

A 5A (BS 3036), rewireable fuse which blows only when 9A flows will have a fusing factor of $\frac{9}{5} = 1.8$

Typical fusing factors

Semi-enclosed fuses: between 1.5 and 2

Cartridge fuses: between 1.25 and 1.75

HBC fuses: less than 1.5

Circuit breaker operation: less than 1.5.

Example:

Determine the minimum fusing current of a:

a   15A BS 3036 rewirable fuse

b   15A BS 88-3 (1361) cartridge fuse.

$$\text{Fusing factor} = \frac{\text{fusing current}}{\text{current rating}}$$

∴ Fusing current = fusing factor × current rating

a   = 2 × 15 = 30A

b   = 1.5 × 15 = 22.5A

From this simple calculation we can clearly see that the BS 88-3 (1361) type fuse offers 'closer protection' than the BS 3036 type fuse, since its fusing current is closer to its actual current rating.

Breaking capacity

The breaking capacity is the maximum current that can safely be interrupted by the fuse.

When a 'short circuit' fault occurs, the current may, for a fraction of a second, reach thousands of amperes. Fuses are designed to break these very high currents without destroying themselves and causing damage to their surroundings by arcing, overheating or the scattering of hot particles.

Note

A circuit breaker must be able to safely 'make' and 'break' a circuit under all conditions (i.e. fault current and normal load current).

Task

Using manufacturers' data, find the breaking capacities for the different types of BS EN 60898 circuit breakers.

_____

_____

_____

_____

Part 2

Fuses

Fuses are the most common type of overcurrent protection device in use. These devices have been around for more than 100 years providing protection against all types of overcurrents.

In basic terms fuses consist of a small diameter wire installed in the circuit. If an excessive amount of current begins to flow in the circuit this piece of wire gets hot. As the current flow increases it gets hotter and hotter until it finally melts and opens the circuit, and we say the fuse has 'blown'.

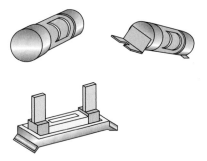

Figure 9.9 _Common types of fuse_

Not all fuses are made in the same way so let's look at the various types of fuses in use.

Semi-enclosed fuse (BS 3036)

This is the good old 'rewireable' fuse of which there are thousands installed throughout the world.

It is known as a semi-enclosed fuse because the fuse element is only partially enclosed between the carrier and the base. When the carrier is removed, the fuse wire may be easily seen, and the wire element can be replaced when necessary.

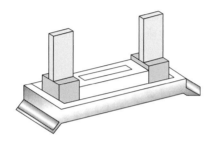

Figure 9.10 *Semi-enclosed fuse with tinned copper fuse wire (BS 3036)*

The main **advantages** of this type of fuse are that:

- they are relatively cheap
- they are easily repaired
- they are fairly reliable
- it is easy to store spare wire
- it is easy to see when a fuse has blown.

These are just some of the reasons why this type of fuse was once the most widely used overcurrent protection device. However there are some limitations, the effects of which have been to reduce the use of this fuse in favour of other types of device.

The main **limitations** are that:

- they are easily abused, the wrong size of fuse wire being fitted accidentally or intentionally
- they have a high fusing factor, they require around twice their rated current to operate, and as a result the cables they protect must have a larger current-carrying capacity

- the precise conditions for operation cannot be easily predicted
- they do not cope well with high short circuit currents
- the wire element can deteriorate over a period of time.

Figure 9.11 *BS 3036 fuse carriers and bases*

Cartridge fuse BS 88-3 (BS1361)

This fuse uses the same principle of a single fuse wire but this time the wire is enclosed in a ceramic or glass body.

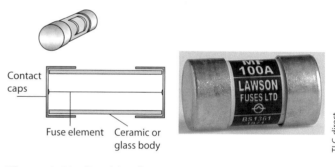

Contact caps

Fuse element   Ceramic or glass body

Figure 9.12 *Cartridge fuse BS 88-3 (1361)*

Because it is enclosed the behaviour of the fuse element under overcurrent conditions can be more accurately predicted.

The main **advantages** of this type are that:

- they have a lower fusing factor, around 1.5 their rated current
- they are less prone to abuse
- being totally enclosed the element does not 'scatter' when it fuses
- they are fairly cheap
- they are easy to replace
- they cope better with short circuit currents.

The main **limitations** are that:

● they are more expensive than BS 3036 fuses
● it is not easy to see if the fuse has blown
● stocks of spare cartridges need to be kept.

This type is a reasonably cheap and more predictable alternative to the rewireable fuse.

When the filament vaporizes, the scattering metal particles are contained within the ceramic body and so present far less of a fire risk than the BS 3036 type.

Figure 9.13 *BS 1362 Plug top cartridge fuse*

The BS 1362 fuse is very similar in construction to the BS 88-3 (1361) fuse but with smaller dimensions to fit plugs and fuse outlets.

High breaking capacity (HBC) cartridge fuse

These fuses are manufactured to BSEN 60269-1:1994 and BS 88 parts 1, 2 and 6.

This is the top of the range with a more sophisticated construction, which makes its operation far more predictable. They may still be known as high rupturing capacity (HRC) types.

The use of a number of shaped silver strips for the fuse element means each individual strip can have a low current-carrying capacity. This in turn gives the fuse a far more accurate operation. Once overcurrent occurs, the first element to 'blow' increases the current flow through the others and so they operate rapidly in an avalanche effect.

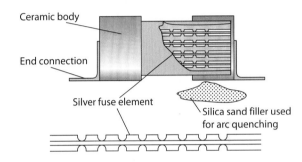

Figure 9.14 *Typical HBC fuse*

The air space within the fuse body is filled with silica sand. This silica sand filler falls into the gap created by the melting elements and extinguishes the arc that is produced. This type of fuse can break short circuit currents in the order of 80 000 A (80kA).

Ceramic body

End connection

Silver fuse element

Silica sand filler used for arc quenching

Figure 9.15 *Typical HBC fuse construction*

The main **advantages** of this type of fuse are that they:

● have a low fusing factor, often less than 1.3 times their rated current
● have the ability to break high currents
● are reliable
● are accurate.

The main **limitations** of the BS 88 part 2 are that:

● they are expensive
● stocks of these as spares are costly and take up space
● care must be taken to replace them with not only the same rating of fuse but with one having the same operating characteristics.

BS 88 gG type fuses are suitable for general applications and BS 88 gM type fuses are suitable for the protection of motor circuits.

Try this

Using a manufacturer's catalogue, list the ratings of the BS 88-3 (1361), BS 1362 and BS 88-2 fuses available.

_____

_____

_____

_____

_____

Part 3

Circuit breakers

Instead of using fuses for protection against overcurrent we can use devices known as circuit breakers. These come in two main categories for internal use:

- circuit breakers (CBs)
- moulded case circuit breakers (MCCBs)

These devices employ a set of contacts which are automatically opened when an overcurrent occurs. This is achieved by using thermal trips for overload and magnetic trips for short circuit. Some devices are available that use only one or the other of these but most now employ both.

Circuit breakers are used in many domestic and commercial installations.

They have the **advantages** of:

- only needing to be reset after operation so no stock of replacements is required
- having settings which cannot usually be altered
- being able to discriminate between harmless transient overloads and still disconnect short circuit faults
- being easy to identify the breaker that has tripped.

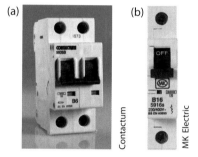

Figure 9.16 _Typical circuit breakers_

The main **limitations** of these devices are that they:

- are expensive
- are mechanical; physically opening the switch to break the current flow
- cannot normally be used if the short circuit current exceeds their short circuit rating.

Using a mechanical switch to open circuits also creates an arc. For lower ratings of CBs the gap created between the contacts may be sufficient to ensure that any arc is extinguished. When operating at higher currents, and where high short circuit currents may be encountered, the circuit breaker must include some method of extinguishing the arc created. Arc splitters are fitted for this purpose.

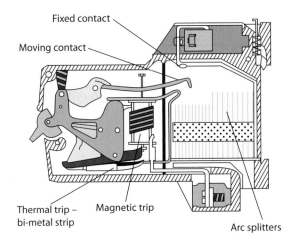

Fixed contact

Moving contact

Thermal trip –
bi-metal strip

Magnetic trip

Arc splitters

Figure 9.17 *Section across a typical circuit breaker*

Figure 9.18 *A typical MCCB EATON 250A, TP*

Discrimination

The objective of discrimination is to make sure that an overcurrent occurring at any point on the electrical system causes the minimum disruption of supply to other circuits and equipment.

Discrimination has been achieved when only the protective device nearest to the fault operates leaving all other protective devices and circuits intact.

For example, in a domestic installation, a fault on a 3kW electrical appliance plugged into a socket outlet on the final ring circuit should result in the BS 1362 13A plug fuse operating while all other final circuits including the ring circuit remain energized.

If the fault resulted in the BS 88-3 (1361), 30A ring final circuit fuse in the consumer unit operating there would be a loss of supply to all other appliances/equipment on that circuit.

Also, if the BS 88-3 (1361), 100A supply authority's service fuse operated, there would be a loss of supply to all final circuits including the lighting circuits and this could present a safety risk and also would be very inconvenient for the consumer.

Discrimination not only applies to fuses and circuit breakers, it also applies to RCDs. Special consideration must be taken into account when selecting different types of protective devices (having different operating characteristics) to ensure effective discrimination is being provided.

Choice of overcurrent protective device

When choosing an overcurrent protective device for a particular application ensure that:

● the correct current rating/type of device is chosen for each part of the system to provide effective discrimination between devices

● disconnection times meet safety requirements to prevent risk of electric shock and damage to property

● the device has a suitable current rating to protect the circuit conductors (the nominal current (I_n) of the protective device must be equal to or less than the minimum current-carrying capacity of the cable (I_t))

● the fuse or circuit breaker can handle surge currents (for example when starting motors).

Try this

Overcurrent protection devices have to be installed in an installation to protect _____, livestock and _____. Fuses to BS 3036 are called semi _____ fuses, fuses to BS 88-3 (1361) are _____ fuses and fuses to BS 88 are _____ cartridge fuses.

The breaking capacity is the _____ current that can safely be _____ by the fuse.

Circuit breakers often have both thermal and magnetic properties, making them suitable to detect _____ currents and _____ currents.

Part 4

Residual current devices (RCDs)

Fuses and circuit breakers may be used to provide protection in the event of a fault to earth. The following devices are specifically designed to protect against earth fault currents.

Residual current device (RCD)

This device senses earth fault currents by measuring current flowing into, and current flowing out of, an installation or circuit and compares the two. If there is a difference between the two currents the 'missing current' must have returned by an alternative route, generally a current flow to earth.

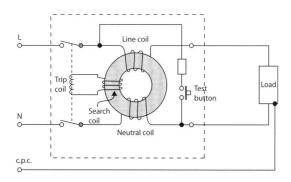

Figure 9.19 *Residual current device*

Figure 9.20 *A typical RCD*

In an RCD there are two main windings wound on an iron core and these carry the **line** and **neutral currents**. Each coil will produce a magnetic field which is directly proportional to the current that flows through it. In normal conditions these two currents will be of the same value and will consequently produce the same amount of magnetic flux.

As the two currents flow in opposite directions, the magnetic fluxes in the ring will cancel each other out and so the magnetic field in the iron ring is zero.

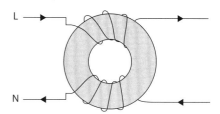

Figure 9.21 *Line and neutral*

If some of the current flows to earth then the line current will be higher than the neutral current.

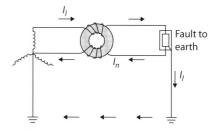

Figure 9.22 $I_l = I_n + I_f$

This means that the magnetic fluxes will no longer cancel out and a residual magnetic flux will circulate in the iron ring.

This residual flux is used to induce a small current into the search coil, and when this reaches a pre-set level the current flowing in this coil is sufficient to operate the trip coil, which in turn releases the powerful spring and a tripping mechanism opens the main contacts isolating the supply.

Test button operation

The test button is connected from the incoming line to the outgoing neutral via a resistor; when the button is depressed the resistor allows a current to bypass the line coil thus causing an imbalance in the core in turn causing the trip to operate.

Residual current devices (RCDs) are often used to provide fault protection in electrical installations as they are able to sense very small leakage currents.

RCBO

The residual current circuit breaker with overload protection (RCBO) combines both RCD protection and CB protection in a single unit and the basic operating principles of this device are the same as those for the individual RCD and CB units previously mentioned.

An RCD may be used to protect a number of final circuits, and if a fault occurs on any one of these final circuits, all of the RCD protected circuits will trip out. This can be very inconvenient for the consumer.

One main advantage of the RCBO is that it will only disconnect the one final circuit it is protecting.

The RCBO has two flying leads, a blue (neutral) lead for connection to the respective neutral terminal and a cream (functional earth) lead for connection to the respective earth terminal inside the consumer unit.

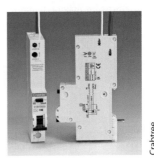

Figure 9.23 *Typical RCBO*

Try this

A _____ (RCD) will not operate when both the _____ and _____ currents are the same, it will operate when there is a _____ current to _____.

An RCBO combines the functions of both an _____ and an _____ in one unit.

Try this

The answers to this crossword may be found in Chapters 7, 8 or 9.

Across

1	Stored energy is called this. (9)
5 & 7	Solid state motor controller. (4, 5)
6 & 4 down	Difference between rotor and synchronous speed. (4, 5)
8	Type of three-phase induction motor rotor. (5)
9	Single-phase motor. (9)
12	Used to improve power factor. (9)
13 & 18	Transformer operating principle. (6, 10)
16	Type of transformer loss. (6)

Down

1	Cell that converts light into electrical energy. (12)
2	0.637 (7, 5)
3	See 15 down.
4	See 6 across.
10	Energy due to motion of an object. (7)
11	A circuit breaker has a thermal and magnetic one. (4)
12	Laminated part of a stator. (4)
14	Electrical variables are measured in these. (5)
15 & 3 down	They repel one another. (4, 5)
17	Earth fault loop _____? (4)

Congratulations you have now completed Chapter 9 of this study book. Correctly complete the self-assessment questions before you progress to Chapter 10.

SELF ASSESSMENT

Circle the correct answers.

1   The most suitable electrical switching device for a 18kW three-phase lighting load would be a:

   a.   double-pole contactor

   b.   solenoid

   c.   triple-pole relay

   d.   triple-pole contactor.

2   When a relay coil is energized, its electromagnetic pole-piece attracts the:

   a.   top moving contact

   b.   top fixed contact

   c.   moving armature

   d.   bottom moving contact.

3   In a BS 88-3 (1361) type fuse the fuse element is contained in:

   a.   asbestos pads

   b.   a ceramic body

   c.   quartz

   d.   air.

4   Arc quenching for a BSEN 60269-1:1994 (BS 88) type fuse is by:

   a.   arc chutes

   b.   silica sand

   c.   air blast

   d.   contact separation.

5   An RCD operates when:

   a.   the line and neutral currents are equal

   b.   the line and neutral currents are not equal

   c.   the sensing coil current is equal to the line current

   d.   the sensing coil current is equal to the line and neutral current.

10 Lighting systems

RECAP

Before you start work on this chapter, complete the exercise below to ensure that you remember what you learned earlier.

1   When current flows through the _____ of an electromagnetic relay, it sets up a _____ field which attracts the _____ to the electromagnet's _____ piece, to _____ and/or _____ contacts.

2   Two advantages of a BS 3036 semi-enclosed type fuse are that they are relatively _____ and they are easy to _____.

3   Two disadvantages of the BS 3036 semi-enclosed fuse are that they do not cope well with _____ currents and they have a _____ fusing factor.

4   A circuit breaker employs a set of contacts which automatically open when an overcurrent occurs. The _____ trip is used when an _____ occurs and a _____ trip is used when a short circuit occurs.

5   An RCBO will operate when an _____, a _____ fault or an _____ fault occurs.

LEARNING OBJECTIVES

On completion of this chapter you should be able to:

● Explain the basic principles of illumination.

● State the applications of the inverse square law, cosine law and lumen method.

● Explain the operating principles of luminaries.

● Identify the different types, applications and limitations of luminaries.

Part 1 Illumination

Electric lighting has been around for many years. In 1809 Humphrey Davy, an English chemist, invented the first electric light, a carbon filament arc lamp (the arc was drawn through the air between two electrodes). In 1878, Joseph Wilson Swan, an English physicist, invented the incandescent-filament electric lamp which also had a carbon filament (the first practical lamp for indoor use). In 1906, the General Electric Company was the first to patent a method of making tungsten-filaments for use in incandescent lamps. Tungsten-filament lamps are still widely used; however, they are far more efficient, safer and brighter than the old type of lamps.

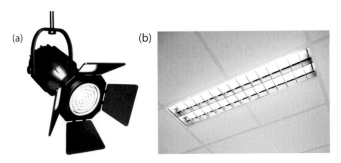

Figure 10.1 *Typical luminaires (light fittings)*

Visible light

Visible light is a very small portion of the complete electromagnetic spectrum, sandwiched between infrared and ultraviolet radiations.

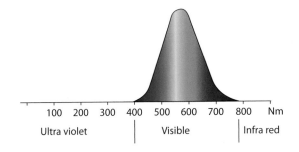

Figure 10.2 *The visible spectrum*

The visible wavelength range, for most people, measured in nanometres, is from about 380nm (violet) to about 780 nm (red).

The various types of light sources use different parts of the visible spectrum and this can have an effect on their application. In areas where red food (red meat in the butcher's shop for example) is on display it is not a very good idea to have a light source that brings out green colours.

Table 10.1

Colour	nm
VIOLET	380–436
BLUE	436–495
GREEN	495–566
YELLOW	566–589
ORANGE	589–627
RED	627–780

A bright yellow lamp is not good if blues and reds are crucial colours, therefore the colour rendering of the light source is a factor that needs to be taken into consideration at the design stage of an electric lighting installation.

Lighting quantities and units

Luminous Intensity (I); unit, candela (cd): this is a measure of the power of a light source (sometimes referred to as brightness).

Luminous Flux (Φ); unit, lumen (lm): this is a measure of the amount of light emitted from a source.

Illuminance (E); unit, lux (lx) or lumens per square metre (lm/m²): this is a measure of the amount of light falling on a surface (also referred to as illumination).

Luminance (L); unit, candela per square metre (cd/m²): this is a measure of the light given off from a surface.

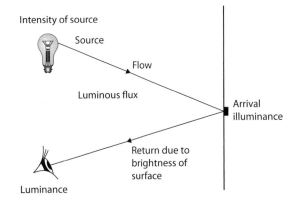

Figure 10.3 *Luminance*

Luminous efficacy (K); unit, lumens per watt (lm/W): this is the ratio of the luminous flux emitted by a lamp to the electrical power input.

Efficacy can be thought of as the 'efficiency' of the light source.

For discharge lighting the lamp power and control gear losses must be taken into account when calculating luminous efficacy.

Luminous efficacy will be used throughout this chapter to compare the different types of lamp and examples of the different lamp efficacies are shown in Table 10.2.

Efficacy is used in place of efficiency because it is very difficult to measure the light output in watts, but very simple in lumens. Lighting manufacturers nowadays also specify the luminous flux output of a lamp in lumens, since it gives the true output of the lamp.

Table 10.2 *Typical lamp efficacies*

Type of lamp	Efficacy (lm/Watt)
(GLS) Filament	10–18
(TH) Tungsten-halogen	12–22
(MPS) HP Mercury	32–58
(MCF) Fluorescent	60–78
(SON) HP Sodium	55–120
(SOX) LP Sodium	70–160

Calculating luminous efficacy

$$\text{Efficacy (lm/W)} = \frac{\text{light output (lm)}}{\text{electrical input (W)}}$$

Example:

Determine the efficacy of a 150W tungsten-filament lamp which gives an average light output of 1960 lumens.

$$\text{Efficacy} = \frac{1960}{150} = 13\text{lm/W}$$

Remember

The input to a lamp is measured in its electrical power unit, the watt, and the light output is measured in lumens.

Sources of lighting from electricity

Whilst there are many different ways of producing light from an electrical source these can generally be divided into three main categories:

1. Incandescent (glowing with heat)

Light is produced by passing a current through a very fine filament wire, and when the wire heats up until it becomes white hot it gives off light.

Figure 10.4 *Incandescent light*

Try this

An 80W tubular fluorescent lamp has an average light output of 5000 lumens, determine the efficacy of the lamp.

_____

_____

2. Discharge (arc)

Light is produced by passing current through a gas or vapour. It is the ionization of the gases or vapour that actually produces the light.

Figure 10.5 *Typical low pressure discharge lights (fluorescents lights)*

Phosphors are often coated on the inside of the lamp envelope to improve light output and colour rendering. The phosphors convert the ultraviolet light emitted from the discharge into visible light.

3. Light emitting diodes (LEDs)

Figure 10.6 *Typical LED lamps*

Light is produced using a diode which produces light when energized.

Reflectors and lenses can be used to further improve light output from all types of lamps.

Try this

Luminous flux is measured in _____ , and illuminance is measured in _____ .

Lumens per watt is the unit of _____ .

The phosphor coating which is applied to the inside of a _____ lamp envelope converts _____ light into _____ light.

Part 2 Inverse square law

The illuminance on a surface which is produced by a single light source varies inversely as the square of the distance from the source. This is known as the inverse square law and provided the light falls at right angles to the surface it can be calculated by the formula:

$$E = \frac{I}{d^2}$$

E = illuminance in lux

I = luminous intensity of the light source in candela

d = distance from the light source to a point on a surface in metres.

The inverse square law is illustrated in Figures 10.7 to 10.9. An incandescent lamp of luminous intensity 500 candelas is fixed at different distances above a flat surface. The value of illuminance E on the surface is calculated for each distance d.

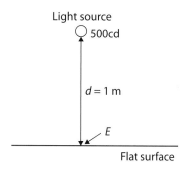

Figure 10.7 *Inverse square law (I)*

$$E = \frac{I}{d^2} = \frac{500}{1^2} = 500 \text{ lux}$$

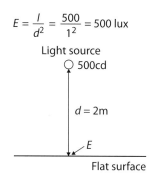

Figure 10.8 *Inverse square law (II)*

$$E = \frac{I}{d^2} = \frac{500}{2^2} = 125 \text{ lux}$$

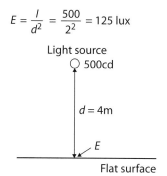

Figure 10.9 *Inverse square law (III)*

$$E = \frac{I}{d^2} = \frac{500}{4^2} = 31.25 \text{ lux}$$

If the distance is doubled between the light source and surface, the illuminance E will be one quarter of the previous value. Although the illuminated area will increase in size, the illuminance on the surface however will decrease accordingly.

Try this

A 1200cd lamp is suspended above a bench. Calculate the illumination at a point directly below the lamp if the height of the lamp above the bench is:

(a) 3m

(b) 5m.

Cosine law

If the light from the lamp falls at an angle to the surface, then at a point on the surface further away from the light source the illuminated area increases and the illuminance decreases. The value of illuminance now depends on the cosine of the angle. This is known as the Cosine Law and can be calculated by the formula:

$$E = \frac{I}{d^2} \times \cos \theta$$

Remember

$$\text{Cos } \theta = \frac{\text{adj}}{\text{hyp}} = \frac{\text{vertical distance 'd'}}{\text{hypotenuse distance 'h'}}$$

Always use distance 'h' in the calculation when using the cosine law formula.

The formula could be expressed as:

$$E = \frac{I}{h^2} \times \text{Cos } \theta$$

The cosine law is illustrated in Figure 10.10. A 500cd incandescent lamp is fixed at a height of 2 metres directly above a long bench, and the value of illuminance at point P is to be determined.

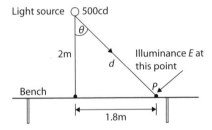

Figure 10.10 *Cosine law*

$$E = \frac{I}{d^2} \times \cos \theta$$

Distance *d* must be found:

$$d = \sqrt{2^2 + 1.8^2}$$
$$= \sqrt{4 + 3.24}$$
$$= \sqrt{7.24}$$
$$= 2.69 \text{ m}$$

Applying the cosine law:

$$E = \frac{500}{2.69^2} \times \frac{2}{2.69}$$
$$= 51.373 \text{ lux}$$
$$\text{illuminance at P}$$

Maintenance factor (M)

When a lamp ages and becomes dirty there will be a loss of light output from the lamp. A maintenance factor is used in lighting calculations to take account of this loss. Maintenance factors in the range 0.8 to 0.9 may be suitable where fairly clean areas are concerned. In dirtier environments, such as welding shops, 0.65 would not be out of place.

Try this

Calculate the illuminance at points P1 and P2 on the working plane.

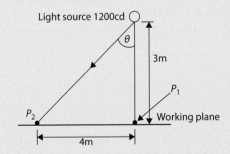

Figure 10.11 *Illuminance*

Coefficient of utilization factor (U)

The coefficient of utilization factor is a measure of the efficiency with which the light emitted from the lamp is used to illuminate the working plane.

The amount of useful light reaching a working plane depends on a number of factors, such as the type of luminaires (light fitting), any reflectors and/or diffusers used, colour and texture of walls/ceilings, room size, number and size of windows, mounting height and position of the luminaires.

Practical values for UF may vary between 0.1 and about 0.95 but 0.1 is a very poor UF value because there is only 10% of emitted light reaching the working plane.

The lighting designer will have to take these factors into consideration in their lighting calculations.

Remember

Not only does the decor of a room deteriorate with age, the lamps and luminaires themselves deteriorate with age (for example the light output from a fluorescent tube gradually becomes dimmer and luminaire reflectors and diffusers gradually become discoloured due to the heat produced by the lamp). Therefore the maintenance factor will be affected by each aspect of deterioration over time.

Lumen method

One important consideration at the design stage of a lighting system is to ensure that the quantity of light reaching a certain surface is of a suitable level. This 'quantity of light' is specified by the illuminance on the surface, and since the level varies across the working plane, an average value is used.

Lighting design guides give typical values of illuminance that are suitable for various rooms and areas. Typical examples of illuminance being: general offices 500 lux, workshops 300 lux and bathrooms 150 lux.

The **lumen method** is used to **determine the number of lamps** that should be installed for a given room or area.

The number of lamps can be calculated using the formula:

$$N = \frac{E \times A}{F \times UF \times MF}$$

Where: N = number of lamps required, E = illuminance level required (lux), A = area at working plane height (m^2), F = average luminous flux from each lamp (lm), UF = utilization factor and MF = maintenance factor.

Example:

A factory production area is 72m long by 20m wide.

If UF is 0.4, MF is 0.75 and the illumination level required for this area is 300 lux, find the number of lamps required if each lamp has a lighting design lumen (LDL) output of 18 000 lumens.

$$N = \frac{E \times A}{F \times UF \times MF} = \frac{300 \times 72 \times 20}{18000 \times 0.4 \times 0.75}$$

$$= \frac{432000}{5400} = 80 \text{ lamps}$$

Space to height ratio (SHR)

Another important factor to be considered at the design stage of a lighting system is the space to height ratio of the luminaires to achieve an even distribution of light.

Try this

A general office 14m by 10m has a design illumination of 500 lux.

If the UF is 0.5, MF is 0.8 and the LDL output of each lamp is 4 400 lumens, calculate the number of lamps required.

_____

_____

Lighting manufacturers give maximum ratios between the spacing (centre to centre) of luminaries and their height (to the lamp centre) above the working plane.

It might be necessary to change the number of luminaries from the original number calculated to achieve an even distribution of light.

Try this

When applying the _____ Law the formula shown below is used,

$$E = \frac{I}{d^2}$$

The cosine law is represented by the expression _____.

Maintenance factor is affected by the deterioration of the _____ itself, and the deterioration of the _____ of the room.

The lumen method is used to find the _____ of _____ required.

Part 3 Incandescent lamps

Lamp designation

The abbreviations used to define types of incandescent lamps are:

GLS general lighting service

TH tungsten-halogen

PAR parabolic aluminized reflector (followed by the lamp nominal diameter in eighths of an inch).

GLS tungsten-filament lamp

Figure 10.12 shows the basic construction of a GLS tungsten-filament lamp (generally referred to as a light bulb).

The tungsten filament is either a single coil or a double coil (coiled coil).

When current is passed through the tungsten filament it heats up and light is given off. This light changes from a red colour at low temperatures to a whiter colour as the temperature rises; however, the light is mainly towards the red end of the visible spectrum which gives an overall warm appearance.

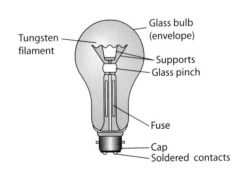

Figure 10.12 *GLS tungsten-filament lamp*

Figure 10.13 *Coiled coil filament*

As tungsten is used as the filament, and this melts at 3 380°C, the operating temperature of the GLS tungsten-filament lamp is kept around 2 500°C. If a tungsten filament was taken up to this temperature in air it would tend to evaporate. There are two main types of filament lamp, vacuum- and gas-filled.

Vacuum-filled type

With this type, the filament operates in a vacuum in the glass bulb (envelope) to reduce the possibility of the filament evaporating. Its efficacy is quite poor and it can only operate at temperatures up to around 2 000°C.

Gas-filled type

The glass bulb is filled with inert gases such as argon and nitrogen (argon to reduce the evaporation process of the filament and nitrogen to minimize the risk of arcing).

The operating temperature is higher than the vacuum type, being 2 500°C, and its efficacy is also better. Using a coiled-coil filament also increases efficacy.

This type of lamp is quite bright; therefore it usually has an opaque coating on the inside of the glass envelope (it is generally known as a 'pearl' lamp).

The tungsten-filament lamps efficacy is only 10 to 18 lumens per watt. As this is low compared with other types of lamp, it is only suitable for situations where high levels of illumination are not required. It is the most familiar type of light source, with an average life of 1000 hours and has numerous advantages, including:

● comparatively low initial cost
● instant light when switched on
● good colour rendering properties
● no control gear required
● can be easily dimmed.

Other types of filament lamps include strip lights, infrared heating lamps, oven lamps, spotlights, floodlights and tungsten-halogen lamps.

Tungsten filaments evaporate unless special conditions are created.

Tungsten-halogen lamps

The linear tungsten-halogen lamp has a small quartz envelope (as shown in Figures 10.14 and 10.15) and with the envelope being very close to the filament, it gets very hot.

quartz can withstand higher temperatures than glass and the pressure inside the lamp can also be considerably increased. This has the effect of slowing down the evaporation of the tungsten and can improve the lamp's life and its efficacy.

Figure 10.14 *Linear tungsten-halogen lamp*

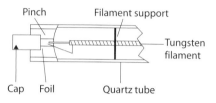

Figure 10.15 *Linear tungsten-halogen lamp detail*

The most important design aspect of a tungsten halogen lamp is the introduction of a small quantity of halogen in the gas filling. The halogen elements used in these lamps include bromine, chlorine and iodine. These halogen elements combine with the tungsten vapour, and when the lamp temperature is high enough, the halogen elements will combine with tungsten atoms as they evaporate and redeposit them on the filament. This recycling process (known as the halogen cycle) increases the life of the filament.

Remember

Tungsten lamps get very hot and need to be kept away from combustible material.

The efficacy of tungsten-halogen lamps ranges from 12 to 22 lumens per watt, which is generally higher than for the GLS tungsten-filament lamp. These lamps light instantly, have a good colour rendering and a warm colour appearance. They have numerous applications including display, photographic, security lighting and floodlighting.

Extra low voltage (ELV) tungsten-halogen lamps

These lamps are produced for use on low (mains supply) voltage systems but also for extra low voltage systems. This has meant that they are used extensively in the automobile industry for vehicle headlamps, as shown in the example in Figure 10.16.

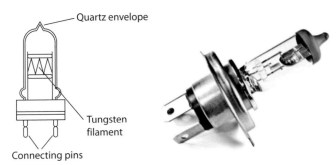

Figure 10.16 *Tungsten halogen (capsule type) lamp and vehicle headlamp*

They can be combined with a precision-faced glass reflector for display spot lights or downlights. The supply for these may be from an inbuilt 230V/12V transformer or a separate 230V/12 V supply.

Figure 10.17 *12V ELV halogen downlight*

Figure 10.18 *12V 35 W MR16 halogen lamp*

Lamp replacement

Contamination, such as grease from finger marks, can considerably shorten the life of a tungsten-halogen lamp. New lamps are packed (with a sleeve) so that they can be installed without touching the quartz. Should they become contaminated they should be cleaned with a spirit before they are used.

Operating positions

Linear lamps should be fitted at or within 10° of horizontal, so that the heat is evenly distributed, however, some low wattage lamps can be operated in any position.

Remember

Incandescent lamps all work on the principle of a continuous tungsten filament that heats up and glows white.

Tungsten-halogen lamps work at very high temperatures. The high bright white light given from the tungsten-halogen lamp makes it ideal for many display applications.

Note

EU Directives have lead to the phasing out of many tungsten-filament lamps particularly the high wattage and poor efficiency types. More information can be found at http://www.parliament.uk/documents/post/postpn351.pdf

Lamp caps/bases

There are many different types of lamp caps and bases, a few commonly used types are listed below:

BC – bayonet cap (B22d has a 22 mm cap diameter with double contacts)

SBC – small bayonet cap

ES – Edison screw

SES – small Edison screw

GES – goliath Edison screw

GU10 – bipin (twist and lock) with 10 mm pin spacing

MR16 – multifaceted reflector, lamp front diameter = 16 × 1/8 inch = 2 inch or 51 mm

MR11 – multifaceted reflector, lamp front diameter = 11 × 1/8 inch = 1 3/8 inch or 35 mm.

Discharge lighting

Discharge lamps rely on the ionization of a gas to produce light and also require some form of control gear to help them strike and then keep them operating efficiently once they have struck. Inductors and capacitors are used for control of discharge lighting circuits.

The inductor (choke or ballast)

The choke or ballast, for discharge lighting circuits, is basically a coil of wire wound round an iron core.

The choke or ballast has two basic functions:

● to initiate the discharge in the lamp, i.e. to cause the electrical arc in the lamp to strike

● to limit the current through the lamp once the arc is struck.

The choke and ballast both work on the principle of electromagnetic induction and when the inductive circuit is broken it causes a high voltage surge across the lamp which is sufficient to strike the main arc in the lamp.

The power factor correction capacitor

A discharge lamp will operate without a power factor correction capacitor; however, the power factor of the circuit would be very poor. To improve this poor power factor a pf correction capacitor is connected across the supply of the luminaire and it is the leading power factor of the capacitor which counteracts the lagging power factor of the inductor (choke or ballast).

Switches for discharge lighting circuits

As we have seen, discharge lighting circuits are inductive and can cause excessive wear on the functional switch contacts (due to contact arcing, mainly when opening the inductive circuit).

When choosing a switch for a discharge lighting circuit, if it is not designed for switching inductive loads, it is recommended that it should be rated at twice the total steady current of the circuit, i.e. a 5A circuit would require a 10A functional switch.

Try this

The glass bulb of a tungsten-filament lamp can be filled with _____. Argon to reduce the _____ of the filament and nitrogen to minimize the risk of _____.

Tungsten-halogen lamps work at _____ temperatures and need to be kept away from _____ materials.

Discharge lamps rely on the ionization of _____ to produce _____.

The choke, which is a type of _____, has two basic functions which are to initiate the _____ in the lamp and to _____ the current through the _____ once the arc is _____.

Part 4 Mercury vapour lamps

There are two types of mercury vapour lamps: Low pressure mercury vapour lamps and high pressure vapour lamps.

Lamp designations

MCF: low pressure mercury - fluorescent lamp

MBF: high pressure mercury with phosphor coating

MBI: high pressure mercury with metallic halides.

Low pressure mercury vapour lamps (fluorescents)

The low pressure mercury vapour lamp, more popularly known as the fluorescent lamp, consists of a glass tube internally coated with a phosphor powder. This tube is filled with mercury vapour at low pressure and a small amount of argon to assist starting. Cathodes that are coated with an electron emitting material are sealed into

each end of the glass tube and connected to the pins of the lamp caps as shown in Figure 10.19.

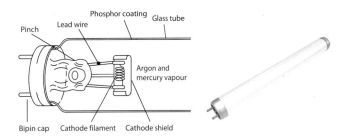

Figure 10.19 *Low pressure discharge tube (fluorescent tube)*

An electrical discharge takes place when a high voltage is applied across the ends of the tube. This electrical discharge produces mostly ultraviolet light. The light output from the tube is produced by the internal phosphor coating when it converts the ultraviolet energy produced by the discharge into visible light.

Control circuits

There are a number of different control circuits, five of which are switch start, electronic start, semi-resonant, lead-lag and high frequency.

Let's look at the operation of the basic switch start circuit.

Switch start circuit

The circuit for the switch start fluorescent is as shown in Figure 10.20.

There is a radio interference suppressor (RIS capacitor) inside the starter switch to cut down interference when striking the lamp.

When the supply is first switched on, current flows through the inductor and lamp filament A, the gas in the starter switch lamp filament B and back to the supply

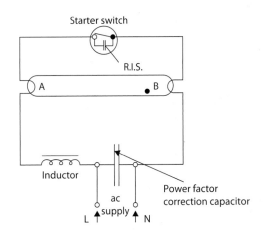

Figure 10.20 *Switch start circuit*

neutral. As the gas in the starter switch heats up, it closes the bimetal contacts. The resistance of the contacts is lower than that of the gas so the gas no longer conducts the current and it cools down causing the bimetal contacts to open again. This momentarily switches off the current flow and a high voltage is induced across the lamp due to the sudden collapse of the magnetic field in the inductor.

The high voltage creates a discharge across the lamp and a new circuit is now completed (via the gas in the tube) effectively shorting out the starter switch. The discharge across the lamp has a very low resistance and if the lamp was connected directly across the supply a very large current would result. The inductor (choke) is now in the circuit to see that this does not happen and limits the amount of current that is allowed to flow. The lamp will now continue to give off light until the circuit is switched off.

The circuit described uses a 'glow' type starter which will need replacing from time to time. This type of circuit has the advantage that these starters are comparatively cheap. There are however several limitations, including:

- access necessary to replace the starter switch
- repeat starting cycles if the lamp does not start (tube flickers)
- when the starter fails repeated starting attempts shorten the life of the other control equipment.

Electronic starters can provide a faster and flicker free start for a lamp and since it has no moving parts there is no chance of mechanical failure.

High frequency circuits operate at about 30kHz and have several advantages over standard ballast unit circuits. These include: fast, first time starting, no stroboscopic effect, higher lamp efficacy, ballast shut down automatically for lamp failure and can be regulated/controlled

It is important that supply cables within this type of luminaire do not run adjacent to leads connected to the ballast output terminals as interference may occur.

High pressure mercury vapour lamp (HPMV)

There are many applications for HPMV lamps such as street lighting, high bay lighting in industrial premises and sports centres. This type of lamp is widely used in industry and for outside lighting where colour rendering is not particularly important.

Figure 10.21 shows the basic construction of a HPMV lamp, and it is inside the quartz arc tube where the high

pressure mercury vapour discharge takes place. This is inside a glass bulb coated with a phosphor which converts ultraviolet radiation from the arc into visible light.

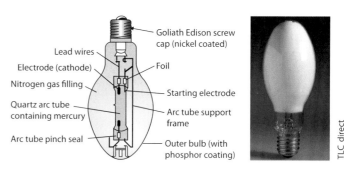

Figure 10.21 *HPMV lamp and 400 W GES MBFU lamp*

An auxiliary (starting) electrode, placed very close to one of the two main electrodes, assists the starting process of HPMV lamps. When first switched on a small discharge occurs between this main electrode and the auxiliary electrode. This causes the main electrode to heat up and trigger the main discharge across the tube between the two main electrodes.

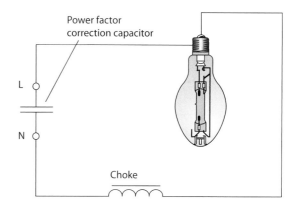

Figure 10.22 *Circuit diagram*

The lamp then builds up energy and in a few minutes produces its full light output. As with fluorescent circuits, the choke limits the current as soon as the arc has struck.

When the supply to the lamp is interrupted it can take several minutes before it is able to return to its full light output. Before the arc can reform again, the temperature of the lamp must drop sufficiently to decrease the internal pressure in the arc tube. Only when the pressure has dropped low enough, will the starting cycle begin again.

High pressure metal halide lamp and control circuit

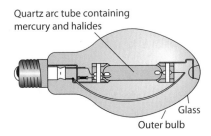

Figure 10.23 *HP metal halide lamp*

Figure 10.24 *400W GES high-intensity metal halide lamp*

Metal halide lamps are very similar in construction to standard high pressure mercury vapour lamps but metal halides (for example: thallium, gallium and scandium) are added to the mercury. The inclusion of these halides improves colour rendering properties and efficacy when compared with the standard HPMV lamps (Figure 10.25).

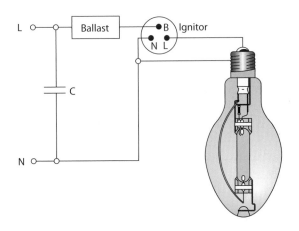

Figure 10.25 *Metal halide lamp circuit*

Metal halide lamps require electrical ballasts to regulate the arc current and deliver the arc voltage. Like HPMV lamps, some metal halide lamps have a third electrode to start the arc when the lamp is first lit, this causes a slight flicker when first turned on. Pulse-start metal halide lamps have an ignitor to generate the high voltage pulse to start the arc (Figure 10.25).

Electronic type ballasts include ignitor and ballast in one single unit.

High pressure metal halide lamp properties: lamp efficacy is between 80–108 lumens per watt, colour rendering is good, lower power consumption than MBF lamps, crisp white light output, average lamp life 15 000 hours. Metal halide lamps should be used in enclosed luminaries.

Limitations: lamp takes time to reach its full brightness, after switching off it will not restart until the pressure inside the lamp has fallen.

Applications: MBI lamps are suitable for any indoor or outdoor commercial and industrial uses where good quality lighting is required, for example, stores, exhibitions, sports stadiums and TV lighting. MBIF type lamps have additional phosphor coatings on the inside surface of the glass bulb to improve the colour rendering properties of the lamp.

Stroboscopic effect

One of the problems with using discharge lighting is what is known as the 'stroboscopic effect'. This effect is created by the flicker of the discharge lamp when it effectively switches off 100 times each second and the current across the discharge tube is continually changing. If this is the only light source over rotating machinery the speed of the rotation of the machine can appear to be slowed down or even stationary even though it is still running at high speeds. This, of course, can be very dangerous, and precautions must be taken to avoid it.

The stroboscopic effect can be reduced by:

- connecting adjacent fluorescent luminaires on different phases
- using special lead/lag fluorescent luminaires
- having a high-power tungsten lamp shining on moving parts
- using fluorescent luminaires operating on high-frequency circuits.

Try this

The low pressure mercury vapour lamp is more commonly known as the _____

A switch start fluorescent circuit has a small capacitor inside the _____, which is for radio _____ suppression, and a large _____ across the supply, which is for power _____ correction.

High pressure mercury vapour lamps will only restrike and return to full light _____, after a _____ failure, when the _____ of the lamp has fallen, which in turn _____ the internal _____ in the _____ tube.

Part 5 Sodium lamps

Lamp designations

SOX: low pressure sodium lamp – single ended

SON: high pressure sodium lamp

SON-T: high pressure sodium lamp – clear, single ended

SON-TD: high pressure sodium lamp – clear, double ended.

Low pressure sodium lamps

This type of lamp has a U-shaped double thickness glass discharge tube with an inner wall made of low-silica glass which can withstand attack by the hot sodium. Inside this tube is a quantity of solid sodium and also a small amount of neon and argon gas to help start the discharge. The outer glass envelope prevents too much heat loss from the inner tube and the two electrodes are of the coiled coil type similar to those in fluorescent tubes. (Figure 10.26)

The recommended burning position of this lamp is ±20° of horizontal to ensure that hot sodium does not collect at one end of the tube to attack and damage it.

Colour rendering is poor since the bright yellow output of the lamp distorts surrounding colours of objects; however, efficacies of over 160 lumens/watt are typical. This type of lamp is normally only used for street lighting and floodlighting.

SOX control gear

The control gear comprises an ignitor and either a leaky reactance transformer or ballast depending on the actual lamp rating (Figures 10.27 and 10.28).

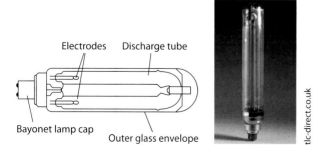

Figure 10.26 *SOX lamp*

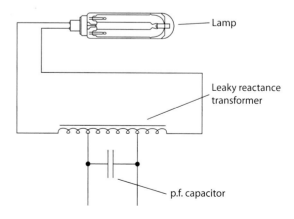

Figure 10.27 *SOX lamps over 100W*

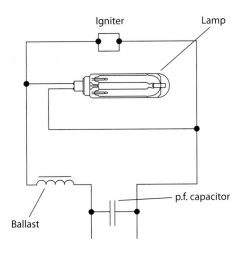

Figure 10.28 *SOX lamps under 100W*

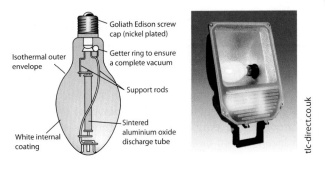

Figure 10.29 *SON lamp and SON floodlight*

Operation: When first switching on a low pressure sodium lamp, a discharge occurs due to the mixture of argon and neon gases (a distinct red neon glow is also produced). The sodium now becomes vaporised by the heat from the discharge and the sodium vapour gradually takes over. This lamp can be restruck when hot within about 1 minute.

High pressure sodium lamps

This type of lamp differs from other discharge lamps having a discharge tube made of sintered aluminium oxide, which is capable of withstanding operating temperatures up to 1 500°C.

The arc tube normally contains sodium doped with mercury and xenon or argon. An outer weather resistant glass bulb, which protects the discharge tube, is completely evacuated to maintain the arc tube temperature at 750°C.

When first switched on, a pulse between 2kV and 4.5kV is required to ionize the argon (or xenon), then as the discharge heats up the sodium vaporizes and takes over the discharge.

This type of lamp can be mounted in any position, has good colour rendering properties and high efficacies of up to 112 lumens/watt, making it ideal for many different inside and outdoor applications (using high bay type luminaries), such as: shopping centres, car parks, sports stadiums and road lighting.

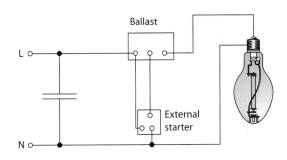

Figure 10.30 *External starter*

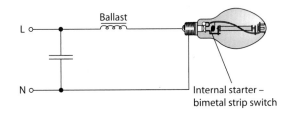

Figure 10.31 *Internal starter*

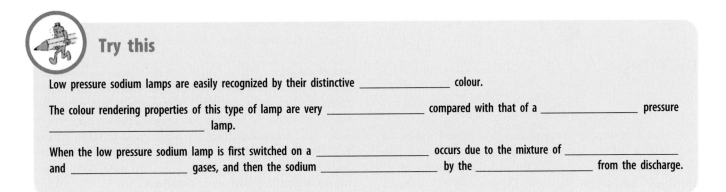

Try this

Low pressure sodium lamps are easily recognized by their distinctive _____ colour.

The colour rendering properties of this type of lamp are very _____ compared with that of a _____ pressure _____ lamp.

When the low pressure sodium lamp is first switched on a _____ occurs due to the mixture of _____ and _____ gases, and then the sodium _____ by the _____ from the discharge.

Part 6 Energy saving, compact fluorescent lamp (CFL)

Compact fluorescent lamps are energy saving lamps and many types are designed to replace standard GLS incandescent lamps. They use about **20 per cent of the power** to produce the equivalent light output as the standard GLS incandescent lamp.

Examples:

An 11 watt CFL will give the same light output as a 60 watt GLS incandescent lamp and a 20W CFL is equivalent to a 100W GLS lamp.

The average rated life span of a CFL lamp is between 10 000 and 15 000 hours, whereas GLS incandescent lamps usually have a lifespan of about 1 000 hours.

There are many different types of CFL lamp, but they all serve the same purpose of reducing the cost of energy bills and lasting longer with increased efficacy, typically in the region of 60 to 72 lumens per watt.

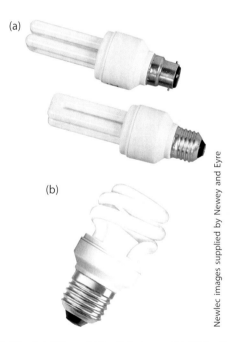

Newlec images supplied by Newey and Eyre

Figure 10.32 *(a) BC and ES stick types (b) ES half-spiral type*

The type of compact lamp shown in Figure 10.33 is an integrated CFL consisting of a fluorescent tube and a ballast control. If the fluorescent tube or ballast fails, the complete unit has to be replaced. Other types of compact fluorescent lamps are of the non-integrated type having separate tubes and control gear.

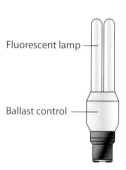

Figure 10.33 *Integrated compact fluorescent lamp*

CFLs cost more than GLS lamps and some types can take a few minutes to attain full brightness after switching on. There are two types of control ballast: magnetic and electronic; the electronic type ballasts are becoming much more popular than the magnetic types which tend to flicker during start up. Electronic types can 'switch on' instantly without any flicker and dimmable types are also available.

The operating principles of compact fluorescent lamps are the same as for standard fluorescent lamps.

Compact fluorescent lamp colours available (similar to standard fluorescents) are warm or soft white, bright white, cool white and daylight. Other colours are also available.

Colour temperature varies depending on the colour of the lamp and is usually indicated in Kelvin's (**K**), for example ≤3 000K is the colour temperature of a warm/soft white lamp and for a daylight lamp it is ≥5 000K. The higher the K value, the brighter the light is.

Colour rendering of a compact fluorescent lamp is usually good; it depends on the phosphor coatings used.

Application: Energy saving lamps (CFLs) can be used for a wide range of domestic, commercial and industrial applications. They are particularly suited for outdoor security lighting which is left on for several hours each night (cold start type), but they are not always suitable for use with motion detectors, especially the lamps with slow 'switch on' times. They are widely used nowadays in restaurants, corridors, toilets and hallways.

Energy saving, LED lamps

Figure 10.34 *LED lamps with GU10 bi-pin (twist and lock mount)*

LED lamps and luminaries are now very popular for many different general lighting and special purpose lighting applications and they have the ability to turn on instantly.

High power 'white' LEDs are used for general lighting systems and various coloured LEDs are used for special purpose lighting systems (especially accent lighting). They have numerous advantages over other types of lamps previously mentioned, such as: very low power consumption, low heat dissipation, no glass tubes to break, resistance to vibration, no minimum current required to sustain operation and, with suitable electronic drivers, are dimmable over a wide range.

LED lamps are more energy efficient than compact fluorescent lamps and offer a lifespan of 30 000 hours or more. An LED light can be expected to last 25 to 30 years under normal use.

Examples: A 7 watt LED lamp gives the same luminous output as a 60 watt incandescent lamp and a 13 watt LED is as bright as a 100 watt incandescent lamp.

A red traffic light signal head that has a cluster of 196 LEDs only consumes 10W, whereas its incandescent counter part consumes 150W; therefore, the LED light consumes 93% less power than the incandescent light.

LEDs are generally too small to be used singly; therefore they are supplied in multiple arrays or modules of different shapes and sizes for general and special lighting applications.

Domestic LED lamps

Domestic LED lamps are interchangeable with normal GLS incandescent lamps having different lamp bases, such as BC, ES and GU 10 bayonet sockets. LED lamps are often supplied from a 230V ac/12V dc LED driver.

LED drivers

LED drivers are self-contained ac/dc power supplies that have outputs that match the electrical characteristics of the LED lamp.

There are two types of LED driver: constant current and constant voltage.

Constant current drivers should be used with constant current lamps and lamps also need to be wired in series, likewise, constant voltage drivers should only be used with constant voltage lamps and they should be wired in parallel.

230 V ac to 12 V dc drivers are used with SELV compliant GU 10 (bayonet) recessed ceiling LED downlights. These are now becoming increasingly popular as a replacement to using either GU 10 Halogen Lamp or GU 10 Compact Fluorescent Lamp downlights.

LED construction

Light emitting diodes (LEDs) are solid-state semiconductor devices and illuminate when the semiconductor crystal is excited so that it directly produces visible light in a desired wavelength (colour) range without having to use coloured filters.

Examples: white, deep blue, blue, green, yellow, amber, orange, red, bright red and deep red.

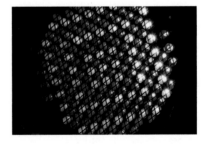

Figure 10.35 *LED array PAR 64 (200mm diameter) theatre stage light*

White light can be produced by mixing red, green and blue LEDs.

Extra bright white light LEDs are made of a semiconductor material composed of indium, gallium and nitride (InGaN). Colour temperature is in the range 4 000–11 000K.

Applications of LED lamps: LEDs are used for many applications such as traffic signals, railway signals, pedestrian crossing signs, street lighting, flood lighting, stage lighting, automobile lights, speed limit and exit signs.

Note

Current building regulations (Part L) states that a certain number of dedicated low energy light fittings must be installed within 'new build' homes and extensions. Approximately 25% of new light fittings should be the fixed energy efficient type.

Try this

Compact fluorescent lamps use about _____ less power than _____ filament lamps and have either a _____ or _____ type ballast.

A CFL with an _____ ballast can switch on instantly without any _____ .

LED lamps are more _____ efficient than _____ fluorescent lamps.

An LED driver is a self-contained _____ power supply.

Congratulations you have now completed Chapter 10. Correctly complete the self-assessment questions before you progress to Chapter 11.

SELF ASSESSMENT

Circle the correct answers.

1 Which of the following statements is *incorrect*?
   a. luminous intensity is measured in Candela
   b. luminous flux is measured in Lumens
   c. efficacy is measured in candela/m^2
   d. illuminance is measured in lux.

2 The highest efficacy lamp is a:
   a. coiled coil tungsten filament
   b. low pressure sodium
   c. low pressure mercury vapour
   d. tungsten-halogen.

3 An 800cd lamp is suspended 2m above a workbench. The illumination at a point directly below the lamp will be:
   a. 1 600 lux
   b. 400 lux
   c. 400 lumens
   d. 200 lux.

4 A compact fluorescent lamp uses less energy than a standard GLS incandescent filament lamp. The approximate energy saving of a CFL compared to an equivalent GLS lamp is:
   a. 98%
   b. 80%
   c. 20%
   d. 10%.

5 The brightest light output will be produced by a:
   a. 2 700K warm white CFL
   b. 2 800K halogen spot light
   c. 4 000K cool white CFL
   d. 6 400K cool white LED lamp.

Electrical heating

11

RECAP

Before you start work on this chapter, complete the exercise below to ensure that you remember what you learned earlier.

1   Calculate the illuminance at points P_1 and P_2 on the working plane.

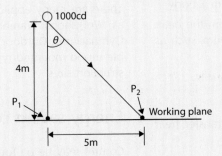

Figure 11.1

Illuminance _____

2   The factors that affect the coefficient of utilization are the luminaire _____ and mounting _____, the _____ of the room, the _____ and _____ of walls and _____ and the number and _____ of windows.

3   In a switch start fluorescent light fitting the _____ provides _____ factor _____, the choke _____ the _____ in the lamp and then _____ the current through the _____ once it is _____. The glow starter when closed heats the tube _____ and when open it open circuits _____ to cause HV _____ across the _____.

4   State three applications where LED lamps are used for general lighting systems.

_____

_____

Part 1 Electric heating

Electric heating is any process in which electrical energy is converted to heat. Two common applications are space heating and water heating.

There are three methods of transferring heat from one medium to another:

● Radiation – where heat is radiated out from a body.
● Convection – where heat is transferred from the heat source by the movement of another medium, such as air or oil.
● Conduction – where heat is transferred usually through a solid medium.

We are mainly concerned with heaters that transfer heat by radiation or convection.

Space heating

Space heating is heating that is used to warm the air in an enclosed area, such as a room, office or workshop.

Radiant heaters

Radiant heaters usually have a coiled tungsten heating element (that reaches very high temperatures) enclosed inside a heat resistant quartz tube with a reflector to direct the heat away from the heater body. The element emits infrared radiation that travels through the air until it hits an absorbing surface, where it is partially converted to heat and partially reflected. This type of heater directly warms people and objects in the room, rather than the air in the room. Quartz halogen infrared radiant heaters are becoming increasingly popular since they are highly efficient, most of the input energy (typically 95%) is converted into heat output, they provide instant heat (which can be directed in a particular direction) and they are very eco-friendly (emitting clean odour free infrared heat).

Applications

There are many applications for this type of heater, such as: Workshops, Warehouses, Churches, Garages, Factories, Gymnasiums, Outdoor smoking and patio areas,etc. Portable tripod mounted types are often used on building/construction sites.

Energy saving controls

Control may be by touch button timer switches or movement sensors (PIR'S) to ensure heaters are not energized when an area is unoccupied.

Limitations

Some types of infrared heaters glow quite bright and produce an uncomfortable glare. Note: Ceramic infrared radiant heaters emit invisible heat without any glare.

Figure 11.2 *Typical infrared radiant heater*

The infrared or halogen type bathroom heaters give out heat by radiation and are often used for supplementary heating.

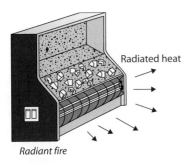

Radiated heat

Radiant fire

Figure 11.3 *Radiant electric fire*

The traditional coal fire operates on the principle of radiation as does the good old radiant electric fire. If a material, such as coal, is burnt then heat is produced. The most common way for this heat to be transferred from the coal to the room is by radiating the heat outwards in much the same way as light is radiated from a light source.

If you have ever stood in front of a coal fire you will also be aware that radiated heat follows very closely the characteristics of light radiation. For example, the further away from the fire you stand the less heat reaches your body and this almost follows the inverse square law as applied to light.

Convector heaters

Convector heaters are those which operate by convection of air currents circulating through the body of a heating appliance, and across its heating elements.

Operating principle:

- cool air flows into the bottom of the heater
- this cool air then passes across the heating elements
- the heated air then rises
- the heated air cools and falls
- the air is reheated again as it circulates through the heater again
- this operation continues until the required temperature is reached (depending on the thermostat setting).

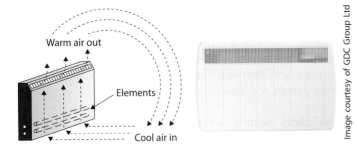

Warm air out

Elements

Cool air in

Image courtesy of GDC Group Ltd

Figure 11.4 *Typical wall-mounted convector heater*

The warm air is constantly circulating in the room, and the more the air is heated, the more the humidity is drawn from it, consequently, making the air drier.

The majority of the heat is transferred to the air by convection although if you are close to the heater you will feel heat being radiated from it.

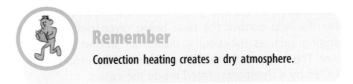

Remember

Convection heating creates a dry atmosphere.

TLC direct

Figure 11.5 *Typical portable convector heater*

Applications

Convector heaters fall into a number of categories so we'll look at some typical types and applications. Low wattage units: these are generally used to maintain a relatively low ambient temperature in areas such as greenhouses, stores, sheds and airing cupboards. These units are usually self-regulating containing inbuilt thermostats and are left on continuously under their own control. Larger, oil filled radiators, are often used to heat bigger areas such as domestic bathrooms. Again, the room temperature is controlled by an integral thermostat within the heater.

Limitations

These units could be used to provide a full space heating system, however they are not very efficient. Whilst they can raise temperatures in larger rooms their make up rate, due to losses from, for example, opening a door, are so slow that such a system would be both inefficient and fairly ineffective.

Fan assisted convector heaters can be used to provide an even distribution of warm air throughout the room.

Storage heaters

Total electric space heating generally makes use of low-cost electricity at night. These systems are controlled by radio tele-switches on special tariffs which charge customers about half price for electricity used over a particular period at night. Such heating devices have to be designed so that heat produced at night can be stored and given off during the day.

Special storage heaters have been developed that consist of elements embedded in refractory bricks as shown in Figure 11.6. The elements heat the bricks up at night and they retain the heat and give it out as required during the day. To help control the heat being given out, thermal lagging such as rockwool is placed all around the brick core. The amount of heat being put into the heater is controlled by a thermostat fitted inside the case.

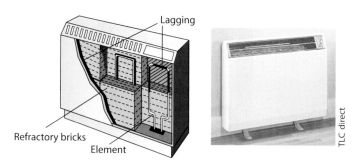

Figure 11.6 *Typical storage heater*

Types of storage heaters

Combination storage heaters combine the benefit of a domestic storage heater charging throughout the low tariff period and an additional convector heater in the same unit which can be thermostatically switched on at any time to provide additional heat when required. Automatic types have an ambient temperature thermostat which optimises the charge to suit room conditions and also save on energy.

Fan assisted storage heaters are more energy efficient than standard storage heaters as they have more thermal insulation lagging and have a low noise, thermostatically controlled fan to ensure the stored heat is delivered only as required. They also provide a greater heat circulation and overall warmth throughout the whole room they are heating.

Automatic charge control storage heaters are also more energy efficient than standard types since they measure the room temperature whilst charging and then store the appropriate amount of heat. An automatic charge regulator adjusts the level of the input charge when weather conditions change.

Note

To comply with Part L of the Building Regulations, all storage heaters in new build properties must include automatic charge control.

One significant limitation with storage heaters is that they emit most of their stored heat during the day and so heat emissions are often considerably less in the evening.

Underfloor heating

Underfloor heating was used by the Romans in their villas; nowadays we can install underfloor electric heating in commercial and domestic properties (for primary heating or additional comfort heating). A special heating cable is installed in the final floor screed as shown in Figure 11.7. This is known as a 'loose wire' cable system and the cable acts as the heating element.

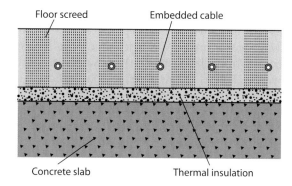

Figure 11.7 *Typical 'loose wire' underfloor heating system*

Three other types of popular underfloor heating systems used nowadays are pre-formed cable mats, wire encased in aluminium foil mats and flat carbon film mats. Once the installation is complete and connected up the whole floor becomes, in effect, a large storage heater. As heat rises the rooms are warmed by natural convection. This is another heating system that can use the off peak facility offered by the electricity supplier.

It is very important to identify areas of floor where fixed equipment is to be located, or where fixtures will need to be screwed to the floor (for example toilet pans and washbasin pedestals), as heating cables must not be installed where these situations occur.

The main advantages of this system are that it is invisible, it does not occupy any space within the finished room and it is maintenance free. One main limitation is that the response time is slow especially when cables are embedded in solid screed floors. We may use this type of system to provide soil warming for horticultural applications or to keep ramps and roadways free from ice.

Underfloor heating systems are widely used in domestic premises, such as under stone or ceramic floor tiles in bathrooms and kitchens, under laminate or wood flooring finishes in living rooms and under carpet.

Try this

Ignoring resistance tolerances, determine the nominal resistance of a 2m^2 underfloor heating mat rated at 230V, 160W/m^2.

_____

Loose wire or pre-formed cable mats are suitable for under stone or ceramic floor tiles, aluminium foil for under laminate or wood flooring and carbon film for under carpets.

Cable mat output power ratings vary depending on their particular application, typical examples are: 120W/m^2, 160W/m^2 or 200W/m^2.

Digital programmable thermostatic controllers are used along with sensitive floor temperature probes, using thermistors, for optimum temperature control of this type of system.

Loose wire cables and pre-formed cable mats have multi-stranded copper conductors, tough double insulation, earth screen to bond to the consumers earthing system and an overall PVC sheath. The overall thickness of cables is usually only 2 or 3mm and therefore, the floor height

will not be raised too much by using these heating systems.

TLC direct

Figure 11.8 _Pre-formed underfloor heating cable mat_

Continuity and insulation resistance tests should be carried out prior to installing cables or mats, immediately after installation and before commissioning the system.

Try this

Infrared _____ heaters usually have a coiled _____ heating element inside a _____ resistant _____ tube.

With a convection heater the _____ air is constantly _____ in the room.

Storage heaters have elements embedded in refractory _____ surrounded by _____ lagging.

Loose wire _____ and pre-formed cable _____ are used for _____ heating.

Part 2 Water heating

Electric water heaters are used in a wide variety of domestic, commercial and industrial applications. With many different heaters available correct selection is important but, generally speaking, these fall into two main categories;

storage and instantaneous types. We shall consider some of the types of heaters available and look at their most suitable applications, mainly in domestic installations.

Storage types

This is one of the most common forms of water heating to be found in a domestic installation and is usually in the form of a large hot water tank fitted with an immersion heater which normally supplies all the hot water taps in the house. A simplified version is shown in Figure 11.9.

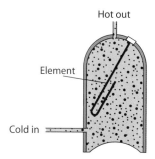

Figure 11.9 *Simple immersion heater*

Storage water systems such as these have the advantage of making a large quantity of hot water available at any one time and usually with a fast delivery rate. They can also take advantage of the off peak electricity tariff. Their main limitations are that once the stored water is used up they have a long re-heat time and they are subject to heat losses when water is not being used. To help overcome these problems some manufacturers have a version that has two immersion heaters as shown in Figure 11.10.

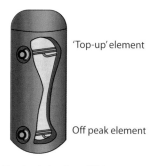

Figure 11.10 *Typical dual tariff immersion heater*

These are usually connected so that the lower element is supplied during off peak times and the whole tank is heated. The top element is only used during high rate times when the tank's hot water has been used. This element tops up the tank until the off peak element comes back on via the time switch.

Instant types

The instantaneous electric shower is a common feature in many homes and operates on the principle of a restricted flow of water through a small tank containing a high power element; 9.5kW and 10.5kW showers are now increasingly popular but they are not very economical to run if they are used for considerable periods of time. However they are often quicker and cheaper than running a bath.

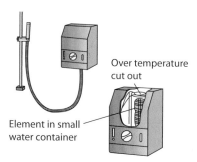

Figure 11.11 *Shower unit*

Figure 11.12 *9kW Mira Shower*

A smaller version is used to provide hot water to sink taps. This type of water heater has the advantages of being economical, as water is only heated when it is required, and is able to supply unlimited amounts of hot water, within reason.

Their principle limitations are that the flow rate is considerably less than that from a storage system, they can usually only supply one tap and in the event of a power failure no hot water is available. Whilst this is also true of the storage types, with the instantaneous types the lack of stored hot water is also immediate and of course they cannot take advantage of the off peak tariff as the storage type does.

Electric heating control

Temperature control

Having considered the various methods of heating using electric current we must consider the control of the heating process. We cannot allow heaters to run unchecked as this would give rise to an unacceptable and in some instances a dangerous situation.

The most common method of temperature control is by the use of a thermostat. This may be an integral part of the heater as in the case of fan heaters, night storage heaters, oil filled radiators and the like. It may also be required as a remote sensor for frost protection systems and heating systems other than those powered electrically, such as gas fired boilers.

The operation of a thermostat generally relies on a change in a material when it is subjected to a temperature change. The simplest example of this is the bimetal strip.

Bimetal strip

In Figure 11.13 we can see a simplified construction of a bimetal strip which comprises two materials which have different rates of expansion. These two materials are joined and fixed together along their length. If heat is applied to these materials the one with the greater expansion rate will expand more for any given rise in temperature. If the two materials were not joined the result would be as shown in Figure 11.14. By fixing the materials together the expansion causes the combined strip to bend as shown in Figure 11.15. Note: Invar and brass are often used for the strips, since brass has a higher expansion rate than invar.

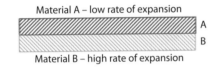

Figure 11.13 *Bimetal strip*

Figure 11.14 *Heat applied to an unbonded strip*

Figure 11.15 *Heat applied to a bonded strip*

If we position a control switch so that it is operated when the bimetal strip bends we can control the source of heat via this switch. If we make the positioning of the switch operating arm variable to the position of the strip we can adjust the switch to allow the strip to bend greater or lesser distances before the switch operates. This will allow

us to vary the temperature change needed to cause the operation of the switch.

A typical room temperature control thermostat is shown in Figure 11.16. A similar device can be used to control a storage radiator. You will see that a small magnet and armature have been added to the assembly. This is to ensure that the operation of the switch is carried out quickly enough to prevent prolonged arcing between the contacts.

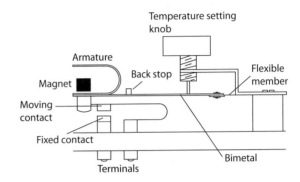

Figure 11.16 *Typical bimetal thermostat*

The rod thermostat

This is the most common form of control for immersion heaters and a typical construction is shown in Figure 11.17. The brass tube and a rod, usually made of invar which has a very low expansion rate, are joined at the tip only. The expansion rate for brass is much higher than that of the invar rod and so as the temperature rises the brass tube expands faster than the rod, effectively pulling the rod into the tube. By locating the switch mechanism so that it is operated by the movement of the rod we can use the device to control the input to the water heater and thus the water temperature.

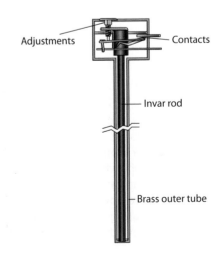

Figure 11.17 *Rod thermostat*

Thermo-couple

A thermo-couple comprises a junction between two dissimilar metals which, when heated, produce a small emf. Whilst this is normally in the region of millivolts it is sufficient to be monitored by an electronic device or a sensitive voltmeter, scaled to indicate temperature as opposed to voltage. We can use these to illustrate temperature or, with the addition of the appropriate electronic relay circuitry, control heating loads.

Variations on these themes are used for all types of heating control and the introduction of heat sensitive electronic components has made the solid state control of shower temperatures and so on commonplace.

These devices are often linked to sophisticated programmed controllers to give a high degree of control within fine tolerances.

Try this

One main limitation with _____ heater cylinders is that they take _____ to reheat once the _____ water is used.

The _____ strip has _____ materials with different rates of _____.

Part 3 Space heating control system

Central heating systems are still widely used for total space heating systems in domestic properties with boilers that burn fossil fuels such as natural gas or oil to heat the water.

The block diagram in Figure 11.18 shows a typical centrally heated space heating system which includes the provision of hot water from the same boiler.

Let's consider the function of each component and device on the diagram.

Input devices

Room thermostat

The room thermostat is required to monitor the air temperature within the space being heated. This will signal for more heat if the air temperature falls below the set level and signal shut off when the required level is achieved. Bimetallic thermostats are still used for domestic systems; however, they are now being superseded by the electronic sensing types using thermistors which respond to temperature changes much faster than the bimetallic

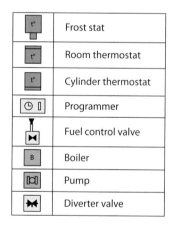

t°	Frost stat
t°	Room thermostat
t°	Cylinder thermostat
⏲	Programmer
⋈	Fuel control valve
B	Boiler
⋈	Pump
⋈	Diverter valve

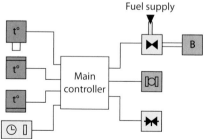

Figure 11.18 *Simple block diagram for a space heating control system*

types. There are different electronic types with digital displays, such as programmable (time and temperature) and also wireless types with remote controls, which enable people to control the central heating from anywhere in the home.

Figure 11.19 *7-day programmable wireless room thermostat*

Cylinder thermostat

The cylinder thermostat monitors the temperature of the domestic hot water. This is usually a bimetallic type and is fitted onto the side of the immersion heater cylinder by means of a cylinder fixing strap.

Figure 11.20 *Hot water cylinder thermostat with bi-metallic sensor*

Frost stat

A frost stat is generally mounted outside a building and acts as an override device calling for heat should the outside temperature fall below the set level. It will generally be used to override any time control to enable the system to maintain a suitable temperature should there be a sudden drop in temperature and will prevent the central heating system pipes and other parts of the system from freezing up.

Note

Modern condensing type central heating boilers are fitted with inbuilt frost protection that normally cuts the boiler in when the temperature falls to 5°C. This only protects the boiler from freezing, not the pipework and the rest of the system.

Pipe stat

The pipe stat controls boiler operation by measuring the return water temperature to the boiler. These usually have a bimetallic temperature sensor. A frost stat is often used in conjunction with a pipe stat installed for energy conservation.

Programmer

The programmer is in effect a glorified time switch, but generally has extra refinements included to allow for various combinations of space/water heating settings including override facilities. It can be programmed for domestic hot water or central heating only or both any time of the day and any day of the week. The majority of programmers used nowadays are electronic with digital displays.

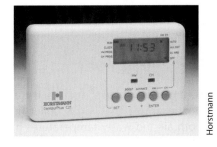

Figure 11.21 *Electronic central heating programmer*

Output devices

There are two main output devices: the pump and the boiler. The function of the pump is to circulate water around the system through the boiler, radiators and hot water cylinder coil. The function of the boiler is to provide heat to raise the temperature of the water being circulated.

Control devices

Main controller

The main controller receives all the data from the input devices and operates all the other control and output devices.

Diverter valve

The diverter valve is operated by the controller to ensure the pumped water is delivered to the priority requirement. This can be set to give priority to the space heating over water heating or vice versa. Three-port mid-position motorised diverter valves are commonly used on fully pumped systems allowing pumped water circulation to both space and water heating when set in the mid position.

Tlc direct

Figure 11.22 *Three-port mid-position motorized diverter valve*

Fuel control valve

This valve controls the flow of fuel to the boiler. If we assume for this example that a gas boiler is being used then when both air and water temperatures have reached the required levels the boiler will shut off. In this event the supply of gas to the burner must also be switched off. The controller carries out this function via the fuel control valve.

Central heating system plans

Central heating systems (controls, pipework and wiring) can be installed in various ways using different system plans.

There are several different types of central heating system plans and details of the operating principles, wiring guides and diagrams for these plans can be viewed on the Honeywell website at www.honeywelluk.com. Two of the most common plans used with fully pumped central heating and domestic hot water systems are the Sundial Y and S plans. Both these plans provide independent temperature control for CH and DHW circuits.

> **Note**
>
> To comply with Building Regulations Part L, central heating systems must have automatic bypass valves and radiator thermostats to prevent wastage of energy from the system.

Both the Y and S plans comply with these regulations; the Y plan has a 3-port mid-position valve, the S plan has 2-port zone valves (automatically controlled by the room and cylinder thermostats in conjunction with the programmer) and thermostatic radiator valves (TRVs) may be fitted in selected rooms/areas as required to improve the energy efficiencies of the systems.

> **Note**
>
> All new central heating boilers should be of the 'highly efficient' condensing type to satisfy part L of the Building Regulations. Conventional and combinational condensing type boilers are commonly installed nowadays since they both have very high efficiencies.

> **Try this**
>
> A _____ thermostat monitors the _____ of the domestic _____ .
>
> The two main output devices of a central heating system are the _____ and the _____ .

Congratulations you have now completed Chapter 11. Correctly complete the self-assessment questions before you progress to Chapter 12.

SELF ASSESSMENT

Circle the correct answers.

1 The most suitable type of heater for a space heating system is a:

a. fan heater

b. infrared wall heater

c. storage heater

d. radiant electric fire.

2 Which type of thermostat is normally used with an immersion heater?

a. bi-metallic

b. rod

c. thermocouple

d. electronic

3 An aluminium foil underfloor heating system would be most suitable for installation under:

a. carpets

b. stone floor tiles

c. ceramic floor tiles

d. laminate flooring

4 The device which diverts water to be pumped through either the central heating or hot water system is a:

a. thermostat

b. bimetal strip

c. programmer

d. motorized valve.

5 Underfloor heating provides warm air in a room by natural:

a. radiation

b. convection

c. conduction

d. induction.

12

Electronic components

RECAP

Before you start work on this chapter, complete the exercise below to ensure that you remember what you learned earlier.

1   Name the three methods of transferring heat from one medium to another

    a _____   b _____   c _____

2   The operating cycle of a convector heater is that _____ air flows into the _____ of heater where it is _____ by _____. It then _____ as warm air, _____ and falls again, re-_____ and _____.

3   State two limitations of immersion heater storage cylinders.

    a _____   b _____

4   Explain briefly the purpose of a cylinder thermostat.

    _____

LEARNING OBJECTIVES

On completion of this chapter you should be able to:

● State the basic operating principles of different types of electronic components.

● Identify typical applications of different types of electronic components.

● Describe the function and application of electronic components that are used in different electrotechnical systems.

● Identify the limitations of electronic components that are used in different electrotechnical systems.

Part 1 Active and passive electronic components

There are two categories of electronic components, active and passive.

All active components or devices are able to control the flow of current through them, such as: transistors, thyristors and triacs. Passive components cannot control the flow of current through them. A fixed resistor is an example of a passive component. It will allow current to pass through it in either direction when a voltage is applied across it. The value of this current depends on the value of the voltage across it. An active device will not allow current to flow through it in either direction. It will effectively act as a conductor one way and act as an insulator the other way, as we will see later.

Passive components consume power and so they often cause a loss of power to the load. Active components can increase the power to the load from a small input signal, for example a transistor, when used as an amplifier.

Resistors

There are many different types of fixed, variable and special purpose resistors. We shall have a look at some of the more common types that are used for general purpose and special applications.

Resistors are connected in circuits for several different reasons, but generally to control voltages or currents.

Resistors used for voltage control are known as potential dividers. Within a circuit it is often necessary to have different voltages at different stages. This can be achieved by using resistors and creating voltage drops.

In the circuit shown in Figure 12.1 resistors x and y have exactly the same value and so the voltage will be divided equally between them.

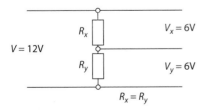

Figure 12.1 *Using resistors to control voltage*

Resistors used for controlling current are known as shunts. Where shunt resistors are used to control currents they have to be connected in parallel with the load.

In Figure 12.2 the circuit current is 0.05A (50mA), but the component indicated as 'a' can only cope with a current of 0.01A (10mA). R_b is connected across it to 'shunt' the current past. The voltage across the component 'a' is 2V, so the same voltage will be across R_b.

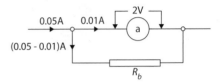

Figure 12.2 *Shunt resistor*

The current through Rb is calculated by:

$$0.05 - 0.01 = 0.04A (40mA)$$

The value of R_b can now be calculated:

$$R_b = \frac{V}{I} = \frac{2}{0.04} = 50\Omega$$

If the voltage is increased the shunt resistor will not offer protection.

Remember

Fixed resistors are passive components which do not have the ability to vary the current flow through them but they are able to control the current or voltage in an electrical or electronic circuit.

Power rating

Resistors often have to carry comparatively large values of current, so they must be capable of doing this without overheating and causing damage. Typical maximum power ratings for resistors are shown in Figure 12.3 on page 212.

Recap the simple power formulae in Chapter 4 before attempting the 'Try this' exercise.

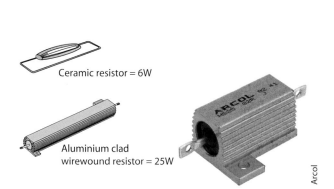

Carbon resistor = 0.5W

Arcol

Ceramic resistor = 6W

Aluminium clad
wirewound resistor = 25W

Arcol

Figure 12.3 *Typical maximum power ratings for resistors*

Equivalent resistance

Series circuit

Unfortunately it is not always possible to have the exact single resistor with the correct combination of resistance and power required. In these circumstances a number of resistors can be connected together.

If a greater value of resistance is required a number of single resistors can be connected in series.

Remember that $R_T = R_1 + R_2 + R_3 + R_4$, etc.

The power of each resistor can be calculated separately once the voltage or current is also known.

Parallel circuit

Calculating equivalent resistors for parallel circuits can be far more complex. In the previous parallel circuit in

Figure 12.2, a resistor of 50Ω carrying a current of 0.04A was required. This 50Ω could be replaced with two equal resisters.

Two resistors of the same value could be used so that the current would be divided equally between them as shown in Figure 12.4. When two identical resistors are connected in parallel the combined total resistance is equal to half of one of the resistors.

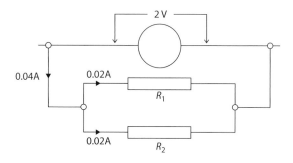

Figure 12.4 *Parallel shunt resistors*

So for two identical resistors in parallel:

$$R_T = \frac{R_1 \times R_2}{R_1 + R_2} = \frac{100 \times 100}{100 + 100} = 50\Omega$$

This means that to end up with a total value of 50Ω each resistor in Figure 12.4 will have to be 100Ω. As the current through each has been reduced to 0.02A the power rating will be $P = VI = 2 \times 0.02 = 0.04W$ (instead of 0.08W).

Remember where there are more than two resistors in a parallel circuit the formula becomes:

$$\frac{1}{R_T} = \frac{1}{R_1} + \frac{1}{R_2} + \frac{1}{R_3} + \frac{1}{R_4} \dots \text{etc.}$$

Try this

Determine a suitable power rating for each of the following resistors:

1   A 470Ω resistor with 10V across it.

2   A 27Ω resistor with 10V across it.

Variable resistors

A large variety of variable resistors are to be found in electronic circuitry. Their function varies from straightforward control to trimming and adjustment during commission and service routines.

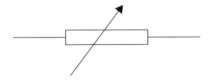

Figure 12.5 *General symbol for variable resistor*

The variable resistor is normally a three-terminal device with connections to either end of a carbon track or wire-wound resistor and a third terminal connected to a sliding contact. In normal industrial practice the change in resistance is proportional to the movement of the slider: this is known as a linear track. In some cases, audio volume controls for example, the track is logarithmic and the change in resistance is not equal to the movement of the slider.

Figure 12.6 *Rotary variable resistor*

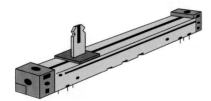

Figure 12.7 *Linear variable resistor*

Figure 12.8 *Symbol for voltage divider with moving contact (potentiometer)*

Thermistors

Thermistors are temperature sensitive resistors.

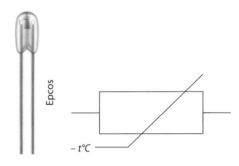

Figure 12.9 *Bead thermistor and NTC thermistor symbol*

Thermistors belong to a group of resistor made from semiconductor materials which are thermally sensitive. They have a controlled temperature coefficient which may be positive (PTC) or negative (NTC). Their uses include temperature measurement and control, temperature stabilization, current surge suppression and a wide variety of other applications. They are suitable for both ac or dc circuits as they are not reactive or polarized.

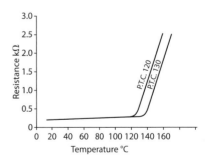

Figure 12.10 *Typical characteristics for 'barium titanite' PTC thermistors*

Resistance is low and relatively constant at low temperatures. The rate of increase becomes very rapid at the switching temperature point. Above this point the characteristic becomes very steep and attains a high resistance value.

Light-dependent resistors

In addition to temperature sensitive resistors there are also light sensitive resistors. These consist of a clear window with a cadmium sulphide film under it. When the light shines on the film its resistance varies. As the light increases the resistance reduces.

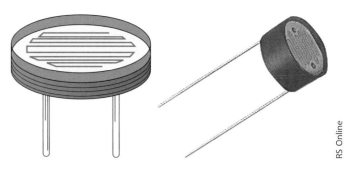

Figure 12.11 *Light sensitive resistor*

RS Online

Try this

Resistors are generally connected in circuits to control _____ and/or _____ .

In order to _____ the current the resistor has to be placed in _____ with the load.

When connecting two resistors in parallel to replace a single resistor each resistor has to have _____ the resistance, but _____ the power rating where they are identical.

Thermistors are _____ control resistors made from _____ materials.

Part 2 Resistor identification

Resistors are normally easy to recognize by their shape and the presence of colour bands or number-letter codes.

Colour code

The colour code for fixed value resistors assigns a numerical value to each one of a range of colours as follows.

Table 12.1 *Resister colour code*

Colour	Value
black	0
brown	1
red	2
orange	3
yellow	4
green	5
blue	6
violet	7
grey	8
white	9

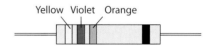

Yellow Violet Orange

Color	band 1	band 2	band 3	band 4
Gold			x 0.1	±5%
Silver			x 0.01	±10%
Black		0	x 1	
Brown	1	1	x 10	±1%
Red	2	2	$x\,10^2$	±2%
Orange	3	3	$x\,10^3$	
Yellow	4	4	$x\,10^4$	
Green	5	5	$x\,10^5$	
Blue	6	6	$x\,10^6$	
Violet	7	7	$x\,10^7$	
Grey	8	8	$x\,10^8$	
White	9	9	$x\,10^9$	

Figure 12.12 *Resistor colour code*

First Band	Yellow	First digit	4
Second Band	Violet	Second digit	7
Third Band	Orange	No. of zeros	3

i.e. 47 000ohms (47kΩ)

Example:

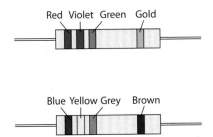

Figure 12.13 *Resistor code example*

Tolerance

The fourth band on the resistor indicates the % tolerance as shown in Table 12.2

A 1 kΩ resistor for example may have a silver fourth band. This, as you can see from Table 12.2, indicates that it has a tolerance of ±10%. This means that its actual value lies

Table 12.2

Colour (4th band)	Tolerance ±
brown	1%
red	2%
gold	5%
silver	10%
no 4th colour band	20%

somewhere between 1 000 + 10% (1 100) and 1 000 − 10% (900). A similar resistor having a gold fourth band could have a value anywhere between 950 ohms and 1 050 ohms, its tolerance being ± 5%.

Where a resistor has three colour bands only, this indicates that it has a tolerance of ± 20%.

Try this

Determine the values of the resistors in Figures 12.14, 12.15 and 12.16.

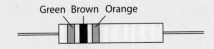

Figure 12.14 *Resistor (I)*

Figure 12.15 *Resistor (II)*

Figure 12.16 *Resistor (III)*

Try this

Using the resistor colour code determine for Figures 12.17 and 12.18:

1 the minimum value of each resistor.

2 the maximum value of each resistor.

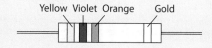

Yellow Violet Orange   Gold

Figure 12.17 *Resistor (IV)*

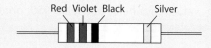

Red Violet Black   Silver

Figure 12.18 *Resistor (V)*

Remember

The first colour band indicates the first digit, the second band indicates the second digit, the third band is the multiplier (no of 0s) and the fourth band is tolerance.

Number-letter code

R = ohms
K = kilohms
M = megohms

So for example:

R56	would mean	0.56Ω
1R0	would mean	1Ω
27R	would mean	27Ω
6K8	would mean	6.8kΩ
470K	would mean	470kΩ
3M3	would mean	3.3MΩ

After this, a second letter is added to indicate tolerance.

F = 1%

G = 2%

J = 5%

K = 10%

M = 20%

and so:

8M2M	= 8.2 megohm	±20%
22 KJ	= 22 kilohms	±5%
4R7K	= 4.7 ohm	±10%

Try this

Using the number/letter code, state the resistance and tolerance for each of the following:

1 10RJ _____

2 1K5K _____

3 1M0M _____

State the number/letter codes for the following values:

1 1.5MΩ ± 5 % _____

2 0.47Ω ± 2 % _____

3 10MΩ ± 1 % _____

Part 3 Capacitors

We have already considered the basic purpose and construction of a capacitor in Chapter 2. There are very few electronic circuits that do not contain a capacitor of some type or other. They can be fixed or variable and range in size from small plastic blobs the size of a solder drip to large cans that look as though they could hold food.

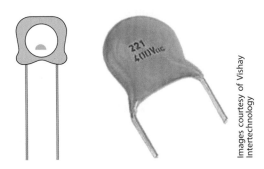

Figure 12.19 *Sub-miniature ceramic capacitor*

Images courtesy of Vishay Intertechnology

Capacitors are used in both ac and dc circuits but their principle of operation is quite different.

Current will not flow through dielectric materials since they are made from insulating materials. The plates of a capacitor are not in direct contact with each other, so

Figure 12.20 *General-purpose electrolytic capacitor and symbol*

Epcos

they don't form a circuit in the same way that conductors and resistors do. However, they do have an effect on the current flow in the external circuit to the capacitor. Let's look at this apparent current flow in dc and ac circuits.

Apparent current flow in capacitors – dc circuit

In the circuit shown in Figure 12.21 the lamp illuminates for a very short period only, this will be the time taken for the capacitor to charge up.

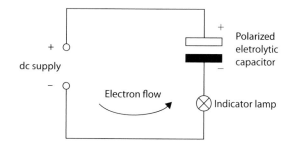

Figure 12.21 *Electron flow*

Once charged, the electron flow will cease and no more movement takes place. Capacitors can be used to block dc current.

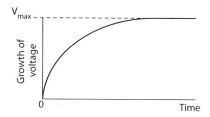

Figure 12.22 *Capacitor charging characteristics (dc circuit) – growth of voltage*

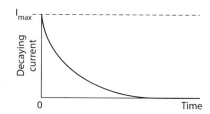

Figure 12.23 *Capacitor charging characteristics (dc circuit) – decaying current*

The graphs in Figures 12.22 and 12.23 show that when a capacitor is charged the voltage gradually rises to a maximum (peak) value and at the same time the current immediately rises to a maximum value and gradually falls to zero. So the current is effectively blocked, since it is zero when the capacitor is charged up to the maximum voltage of the supply.

The maximum voltage that the capacitor will charge up to is that of the supply. On discharge a similar pattern of voltage occurs but going from maximum to zero.

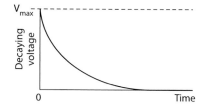

Figure 12.24 *Voltage decay*

Apparent current flow in capacitors – ac circuit

In the circuit shown below the lamp will be illuminated continuously giving the impression that there is a current flow through the capacitor. In fact there is not. When a capacitor is connected, as shown in the circuit, to an ac supply on one half cycle the capacitor will charge up to a particular polarity. On the second half cycle the capacitor will charge up in the reverse polarity to the first half cycle.

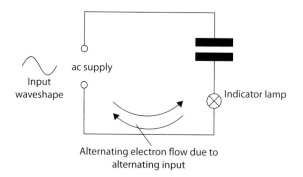

Figure 12.25 *Electron flow*

The electron movement around the circuit will alternate giving the impression that there is a current flow through the capacitor, but of course there is not.

The general rule is that capacitors block dc but pass ac.

Equivalent capacitance

Where it is not possible to replace a capacitor with one of the same value, several can be connected together to give an equivalent value.

Parallel: When capacitors are connected in parallel the total capacitance increases.

$C_{Total} = C_1 + C_2 + C_3 + C_4$... etc.

Series: When capacitors are connected in **series** the total capacitance decreases.

$$\frac{1}{C_{Total}} = \frac{1}{C_1} + \frac{1}{C_2} + \frac{1}{C_3} + \frac{1}{C_4} \dots \text{ etc.}$$

If there are just two capacitors connected in series, the product over sum method can be used to find the total capacitance.

$$C_T = \frac{C_1 \times C_2}{C_1 + C_2}$$

Remember

Capacitors are the opposite formulae to resistors when connected in series and parallel.

Types of capacitor

To a large extent the application determines the type of capacitor that can be used. As with resistors, there are fixed and variable types of capacitor.

Fixed capacitors can be placed into three general classes related to their dielectrics.

Low loss, high stability	– mica
	– low K ceramic
	– polystyrene
Medium loss, medium stability	– paper
	– plastic film
	– high K ceramic
Polarized capacitors	– electrolytic
	– tantalum

Electrolytic capacitors

These capacitors consist of two aluminium foils separated by an absorbent paper. Connections are fitted to the two foils and the whole assembly is rolled up and fitted tightly into an aluminium container which is hermetically sealed. The dielectric is formed electrolytically on the surface of one aluminium foil in the form of aluminium oxide. This foil acts as the positive plate, or anode, of the capacitor. The second plate, the electrolyte, is in contact with the other foil.

As the dielectric is formed by passing a direct current supply through the capacitor it means that the capacitor must always be connected to the same polarity. If it is

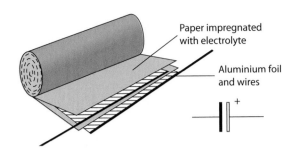

Figure 12.26 *Electrolytic capacitor and symbol*

connected around the wrong way or to an ac supply, the dielectric would break down and the capacitor would become a short circuit.

Non-polarized electrolytics

These capacitors are constructed by using several foils wound into one unit and connected back-to-back. They are basically two electrolytic capacitors wound together in one container. This makes it possible to use electrolytic capacitors on ac supplies.

The construction of these means that they are much larger than ordinary electrolytic capacitors of the same value.

Figure 12.27 *Non-polarized electrolytic capacitor*

Tantalum capacitors

The dielectric in these capacitors is tantalum oxide. This is a much better dielectric than aluminium oxide and gives high values of capacitance in a relatively small space. The construction of these means they are polarized and must be connected in circuits to the correct polarity.

Task

Using manufacturers' catalogues or websites find typical applications for the low loss, high stability and medium loss, medium stability capacitors listed above.

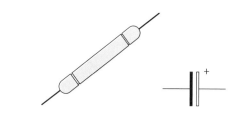

Figure 12.28 *Tantalum capacitor*

Variable capacitors

Variable capacitors consist of one set of fixed plates and one set of moving plates. The greater the overlap of facing plates the greater the value of capacitance.

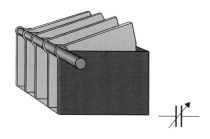

Figure 12.29 *Variable capacitor and the symbol for a variable capacitor*

Identification of capacitors

To identify a capacitor the following details must be known:

- the capacitance
- the working voltage
- the type of construction
- the polarity (if any).

The identification of capacitors is not easy because of the wide variation in shapes and sizes.

In the majority of instances the capacitance will be printed on the body of the capacitor. This often serves as a positive identification of the device as a capacitor.

As we have seen, the value is normally given in microfarads (µF), picofarads (pF) or, less commonly, in nanofarads (nF).

Some capacitors have colour code bands similar to resistors.

The maximum working voltage of a capacitor is normally marked on the side of a capacitor. If this value is exceeded the dielectric would break down and the capacitor would be useless.

Polarity

As we have seen, some capacitors are constructed in such a way that if the device is operated with the wrong polarity its properties as a capacitor will be destroyed.

This is particularly true of electrolytic types.

The positive terminal must never be allowed to go negative, whether by wrong connection, supply reversal or by connection to an alternating voltage.

Polarity may be indicated by a + or – as appropriate. Electrolytic capacitors contained in metal cans may use the can as a negative connection. If no other marking is indicated but if it is still suspected that the capacitor is polarized, a slight indentation in the case will indicate the positive end.

Indent to indicate positive end

Figure 12.30 *Positive indent marking*

Try this

An 8µF and a 12µF capacitor are connected in series. Determine their equivalent capacitance.

_____

_____

Part 4 Semiconductor devices

Semiconductor devices such as diodes, transistors, thyristors and integrated circuits are manufactured from semiconductor materials whose resistivity lies between that of a good conductor and that of a perfect insulator. Silicon is the most common semiconductor material used with added impurities, such as aluminium for a 'P' type material and arsenic for an 'N' type material.

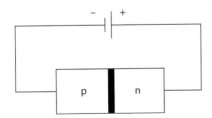

Figure 12.31 *Typical semiconductor devices*

PN junction diode

A PN junction diode is basically a piece of p-type and a piece of n-type semiconductor material joined together.

If a dc supply is connected across a diode, as in Figure 12.32 (reverse biased), the diode will act as an insulator.

Figure 12.32 *Reverse biased*

If the dc supply is now changed, as in Figure 12.33, the diode is forward biased and acts as a conductor.

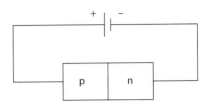

Figure 12.33 *Forward biased*

A silicon diode will conduct when the forward voltage drop is about 0.7V (Figure 12.34) and if the maximum reverse voltage (about 1 200V in this example) is exceeded the diode will break down, conduct and be destroyed. Likewise, the diode will break down if the current rating is exceeded, since excessive heat will be generated.

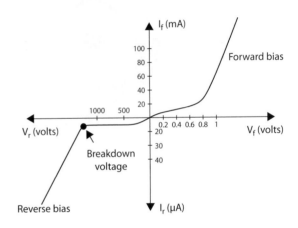

Figure 12.34 *Silicon PN junction diode*

When a single diode is connected to an ac supply (Figure 12.35) it will only allow the current to flow when the p-type is more positive than the n-type. This means that only one half of the current will be conducted. It is in effect switching off the other half of the waveform. This is known as half-wave rectification.

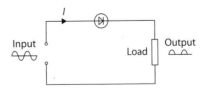

Figure 12.35 *Half-wave rectification*

Figure 12.36 *General symbol for PN diode*

It is important that diodes are correctly installed and for this purpose it is necessary to ensure that the leads or terminals have been correctly identified (Figures 12.37).

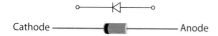

Cathode —————————————— Anode

Figure 12.37 *Direction of current flow is anode to cathode*

Full-wave rectifier units frequently make use of encapsulated bridge rectifiers which incorporate four diodes in a bridge configuration. These have four terminals: two ac input and two dc output, clearly marked + and −.

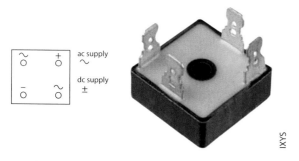

Figure 12.38 *Encapsulated bridge*

Diodes in common use

Power diodes

These are diodes in general use which carry currents of 1A and above.

Signal diodes

Signal diodes are small in size and have limited power-handling capability. They are used in high-speed switching circuits or high-frequency communications circuits.

Zener diode

An ordinary semiconductor diode will only withstand a certain amount of reversed voltage before it breaks down and current flows. Once it has broken down it is permanently damaged. A Zener diode is designed to break down without damage at a predetermined voltage and is usually used as a voltage reference. It can be used

to provide a constant voltage even when the supply voltage and load current may vary.

It is this characteristic of the Zener diode which makes it ideal for stabilizing or regulating the output of power supplies.

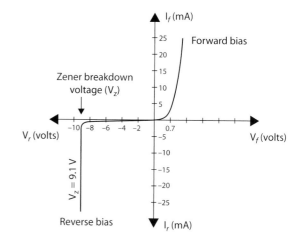

Figure 12.39 *Forward and reverse bias characteristics of a 9.1V Zener diode*

Note

The forward bias characteristic is just like the ordinary silicon P-N junction diode.

Preferred values of Zener diodes are manufactured, for example, 2.7, 4.7, 5.1, 6.2, 6.8, 9.1, 10V, etc.

In Figure 12.40 a 3V load is supplied even though the 6V supply may fluctuate slightly. So that the Zener diode is not damaged a current limiting resistor (R) is used.

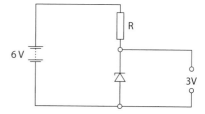

Figure 12.40 *Zener diode voltage regulation*

Figure 12.41 *Zener diode symbol*

Dc supplies from an ac source

Ac to dc converters

There are several different circuits that can be used to pro-duce a direct current from an ac source. Usually the circuit depends on the degree of smoothness of the dc that is required. If the dc is not very critical then a simple half wave circuit can be used as shown in Figure 12.42. It has been assumed in all of the diagrams that the dc voltage is less than the ac and has to be transformed down.

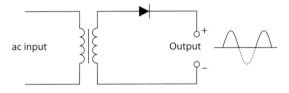

Figure 12.42 *Half-wave rectifier circuit*

This can be made a smoother output by placing an elec-trolytic capacitor across the load terminals.

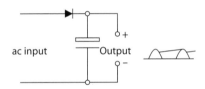

Figure 12.43 *Capacitor smoothing*

Where a better output is required a full-wave circuit is used. This may use a full-wave bridge circuit or a transformer with a centre tap on the output and a bi-phase circuit.

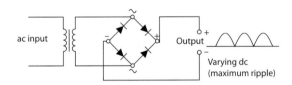

Figure 12.44 *Full-wave bridge rectifier circuit*

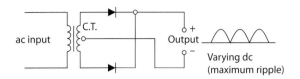

Figure 12.45 *Bi-phase (full-wave) rectifier circuit*

These circuits can be smoothed out if required by using some form of filter circuit.

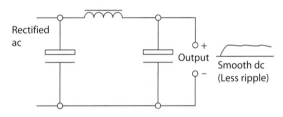

Figure 12.46 *L and C smoothing filter commonly known as π (pi) type*

The above circuits are ideal if the output is not critical.

Voltage stabilization

The dc output voltage can be stabilized by employing a Zener diode as shown in Figure 12.47.

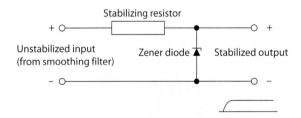

Figure 12.47 *Simple voltage stabilization circuit*

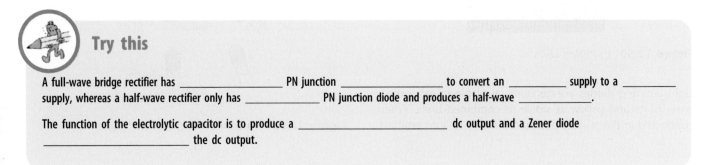

Try this

A full-wave bridge rectifier has _____ PN junction _____ to convert an _____ supply to a _____ supply, whereas a half-wave rectifier only has _____ PN junction diode and produces a half-wave _____.

The function of the electrolytic capacitor is to produce a _____ dc output and a Zener diode _____ the dc output.

Photodiode

A photodiode is basically a PN junction which is enclosed in a can with a window. When light shines on the junction, electrons and holes are produced and current can flow until the light is cut off. This makes them particularly suitable for counting and monitoring processes.

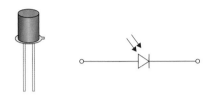

Figure 12.48 *Photodiode and symbol*

Light-emitting diode (LED)

This is basically another junction diode, but when it is connected in the forward direction, light is given off. The light comes from the junction itself, which is placed close to the surface of the diode and in many cases has a lens built above it to give the maximum output. On single units the cathode is usually indicated by a flat on body and a shorter lead.

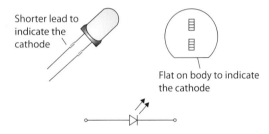

Shorter lead to indicate the cathode

Flat on body to indicate the cathode

Figure 12.49 *LED and symbol*

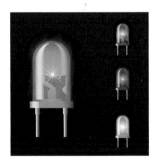

Figure 12.50 *Coloured LEDs*

Light-emitting diodes are generally available in red, green, blue, white and yellow as single or combined units. When connected in the reverse direction there is no light emitted.

Transistors

The transistor is a three-terminal device which is capable of amplifying current. A single bipolar transistor has three connections the emitter, base and collector. The bipolar transistor can be used as a device where a signal current of relatively small proportions can be fed into the base. This will produce a larger current in the collector circuit which can be used for a higher power application. The collector current will only flow as long as the base current is present.

Figure 12.51 shows a circuit where a small input signal controls the operating coil of a relay.

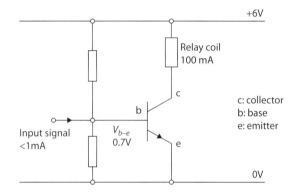

Figure 12.51 *Trigger circuit*

The identification of the three connections depends on the type of enclosure and mounting arrangement used. Figure 12.52 shows some examples, but the exact configuration for a particular transistor should always be checked with the manufacturer's data before it is connected into the circuit.

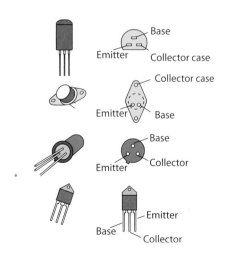

Figure 12.52 *Pin connections vary with type*

There are two main types of transistor: pnp and npn.

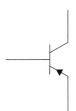

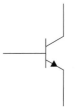

Figure 12.54 *Npn symbol*

Figure 12.53 *Pnp symbol*

Try this

The cathode of an LED is identified by the _____ on the side of its body and the _____ lead.

The three connections of a bipolar transistor are:

a) _____ b) _____ and c) _____

Part 5 Thyristor

A thyristor is a type of silicon-controlled rectifier (SCR) which has cathode, anode and gate connections. The gate connection is usually smaller than the other two and the anode connection is often a stud fixing to a heat sink. A heat sink dissipates any heat generated within the device when it is handling high currents, for example the stud-mounted thyristor in Figure 12.56 below can switch up to 230A and a gate current of about 150mA is sufficient to trigger it into conduction.

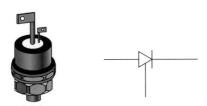

Figure 12.55 *Typical thyristor and symbol*

Images courtesy of Vishay Intertechnology

Figure 12.56 *230A 1 200V stud thyristor*

Thyristors are used in switching circuits. They are four-layer semiconductor devices which do not conduct until a short trigger pulse is applied to the gate and they will continue to conduct after the gate pulse has been removed (Figure 12.57).

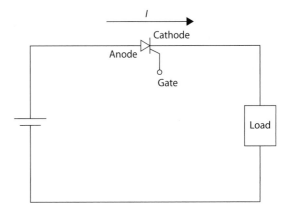

Figure 12.57 *Typical thyristor circuit*

The thyristor is a unidirectional (dc) device and will only conduct in the direction indicated (anode to cathode) like the diode. If connected to an ac supply it will cease conducting after the supply current has reached zero and will only switch on again if the gate pulse is reintroduced in the next or some subsequent positive half cycle. It will not conduct during negative half cycles of the ac supply. Figure 12.58 shows a simple thyristor-controlled ac circuit for a purely resistive load.

Figure 12.58 *Simple thyristor circuit*

The most common application of a thyristor is to control the power supply to a load and it does this by controlling the amount of current it allows to flow through the load during part of each ac cycle. Load power is reduced by triggering the gate later in each cycle. Figure 12.59 shows the chopped current waveform produced due to a delayed trigger pulse at the gate.

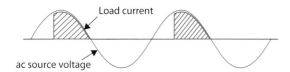

Figure 12.59 *Delayed gate trigger pulse – less than half-wave current through load – less than half power available*

The main limitation with the thyristor is that it allows control of only half the available power in an ac circuit, making it very uneconomical.

The circuit shown in Figure 12.60 includes two thyristors, SCR1 and SCR2, connected back-to-back. This then allows for connection to ac circuits so that one of the thyristors will operate on each half cycle. The main advantage of this arrangement is that the available power to the load can be varied between zero and full load, the same as with a triac.

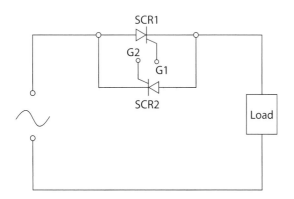

Figure 12.60 *Back-to-back thyristors*

Triac

A triac is a bi-directional switching device which can be used in ac circuits. Triacs are found in motor speed control circuits and light dimmers. Their case and connection layout often get them confused with transistors, as they can look very similar. Triacs are, however, equivalent to a pair of back-to-back thyristors in a single case. A typical configuration is shown in Figure 12.61.

Figure 12.61 *Typical triac and general symbol*

The triac can be switched on (triggered into conduction) in both the positive and negative halves of the ac cycle when a trigger pulse is applied to the gate in each half cycle. The load current and the load power depend on the firing angle of the trigger pulse to the gate. Triggered at:

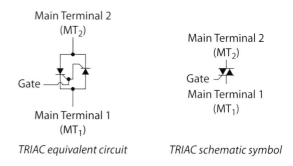

TRIAC equivalent circuit   *TRIAC schematic symbol*

Figure 12.62 *Triac circuit and symbol*

- 0° and 180° – max load current and power
- 90° and 270° – half load current and power
- after 90° and 270° – less load current and power
- 180° and 360° – triac switched off (if not triggered)

Remember

If conduction commences early in each conducting half-cycle, average current and power to the load are much higher than if conduction is delayed.

Figure 12.63 below shows typical waveforms for a purely resistive load with delayed firing angles.

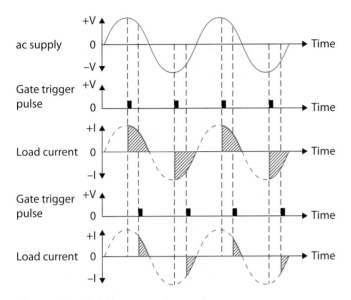

Figure 12.63 *Triac control waveforms*

Note

Delayed firing angles means less conduction.

Diac

A diac is a two-terminal device which conducts in either direction, but only after the voltage across them has reached the level required for conduction to take place. It has only two connections and can be confused with a diode if not careful. Diacs are often used in conjunction with triacs as a triggering device.

Figure 12.64 *Diac and symbol*

Inverters

Inverters are solid state electronic devices that convert direct current to alternating current. The inverter output can be at any required voltage and frequency with the use of appropriate transformers, switching and control circuits. They normally use semiconductor devices, transistors or thyristors to switch the dc input to an ac output.

Simple operation of a single-phase inverter

In Figure 12.65 the dc input (DC Link) to the inverter is from the battery and the ac output from the inverter to the load is taken from terminals A and B.

When transistors TR1 and TR2 are switched ON, the output voltage to the load is positive, and equal to the dc link voltage, while when TR2 and TR3 are ON it is negative. If no transistors are switched ON, the output voltage is zero.

A three-phase output can be obtained by using six transistors.

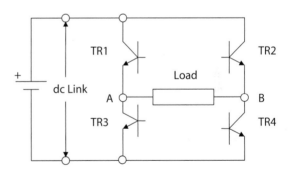

Figure 12.65 *Simplified diagram of inverter circuit for a single-phase output.*

Typical output voltage waveforms from the inverter for a purely resistive load are shown in Figure 12.66.

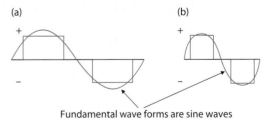

Figure 12.66 *(a) At low frequency; (b) At high frequency*

It can clearly be seen that the resultant output voltage waveform is composed of rectangular or square wave chunks, and is thus far from ideal; however, an ac induction motor can operate from this inferior waveform. Ideally the output waveform should be a pure sine wave.

There are many makes of modified square wave power inverters and pure sine wave power inverters on the market which are very complex in design and are therefore beyond the scope of this book.

Applications of inverters

There are many applications for inverters, a few are:

Dc to ac power supplies: ideal for powering 230V mains equipment from an extra low voltage dc supply, for example, a 12V battery (Figure 12.67).

Uninterruptible power supplies (UPS): take over when mains power is not available.

Variable frequency drives: allowing induction motors to drive machines at very high speeds, for example a CNC Router at 24 000rpm.

Induction heating: operating at high frequency.

Electric vehicle drives: e.g. traction motors in electric locomotives.

TLC direct

Figure 12.67 *Soft-start 12V dc to 230V ac 300W inverter*

Remember

When the inverter is switched ON, it is continually switching the direct current (dc) at the input side (from the dc link) into a continually changing alternating current (ac) on the output side of the inverter to the load.

Try this

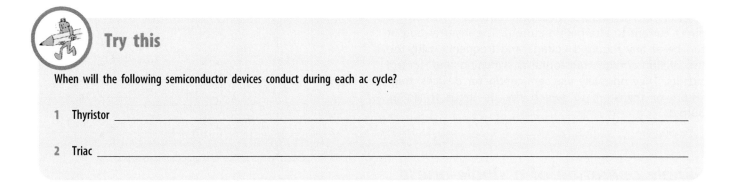

When will the following semiconductor devices conduct during each ac cycle?

1 Thyristor _____

2 Triac _____

Part 6 Application and function of electronic components in electrotechnical systems

Nearly all electrotechnical systems have electronic components soldered directly onto printed circuit boards (PCBs), either in the form of discrete (individual) components or encased in integrated circuit (IC) packages which contain thousands or even millions of electronic components.

For these systems to function they will have to have a suitable power supply, often an ac to dc extra low voltage power supply will be used which will have components such as diodes and capacitors etc., to provide a steady dc output.

Figure 12.68 *PCB with IC and discrete components*

Figure 12.69 *Integrated circuit chips*

We will have a look at a few discrete electronic components which are at the heart of some of these systems.

Dimmer switches

Older type dimmer switches used variable resistors to reduce the voltage drop across a lamp to reduce its power and light output.

Modern dimmer switches use triacs to vary the power and light output.

Remember that the triac can be switched on in both the positive and negative halves of an ac cycle, whereas a thyristor cannot.

The light output from a lamp depends on when (at what firing angle) the triac is triggered into conduction. The earlier the triac is triggered, the brighter the light output, the later the triac is triggered, the dimmer the light output.

Heating control systems

Thyristors are often used to control the output power of electrical heating loads, the earlier the thyristor is triggered in each cycle, the greater the heat output, the later the thyristor is triggered the less the heat output. Triacs can also be used.

Instead of turning thyristors or triacs on and off during each cycle (using phase control triggering) they can be turned on and left on for a number of complete cycles and then turned off at regular intervals, this is known as burst triggering and is a more efficient means of power control for electric heating loads.

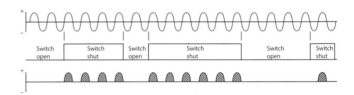

Figure 12.70   *Burst triggering control using a thyristor: top waveform ac supply; middle waveform (square) shows thyristor switch operation; the bottom waveform is load current*

Central heating boiler control systems

There have been numerous changes in the design of CH boiler control systems. At the heart of the system is the PCB control unit, which if it develops a fault and cannot be repaired, is expensive to replace.

Thermistors play a big part in the correct and safe operation of CH boiler systems.

Modern central heating boilers use thermistors to measure, control and regulate the temperature of both the central heating water flow temperature and the domestic hot water temperature. They are also used for over temperature protection of the boiler.

Task

Identify from CH boiler manufacturers data/drawings where thermistors are used.

Motor temperature protection control

Positive temperature coefficient (PTC) type thermistors are widely used to prevent damage to motor windings from overheating since their resistance changes very rapidly at a critical temperature (see Figure 12.10). It is this feature of a sudden change in resistance which is used to initiate the switching off of a motor.

A normal three-phase motor has three thermistors connected in series embedded in each phase winding of the stator.

As they carry only very small currents (mA) thermistors are used in conjunction with an amplifier or relay unit for the control of a motor.

Motor speed control

A thyristor or a triac is often used to control the amount of current flowing (in each ac cycle) to control the speed of a small motor, such as a universal motor in an electric hand drill or electric jigsaw.

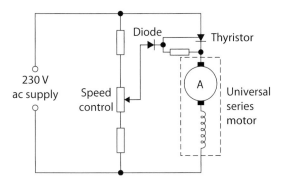

Figure 12.71 *Basic speed control circuit for a small motor*

Security alarm systems

Light emitting diodes (LEDs) are used with active infrared beam motion sensors. These have two units, a transmitter and a receiver. The transmitter has an LED that emits an infrared light beam which is picked up by the receiver that converts it into a dc signal. This dc signal usually operates a relay connected to an audible alarm. When the intruder breaks the infrared beam the signal is broken, the relay de-energizes and the alarm is sounded.

Telephone systems

There have been many changes made to the operation of telephone systems, especially with the introduction of wireless digital cellular (mobile) phone and videophone systems. The traditional analogue landline telephone systems, using electromechanical relay switches, have mainly been replaced by digital systems using solid state switching devices (e.g. transistors, thyristors) for example when connecting to different telephone lines at the telephone exchange.

One major change in modern landline telephone systems is the ability to connect your computer to the internet via the same telephone line as you use for your telephone calls.

Asymmetrical digital subscriber line (ADSL) internet connections are a common type of broadband internet connection which allows users to access the internet and make telephone calls at the same time. ADSL filters (splitters) using inductors, capacitors and resistors prevent interference on the line. They remove lower frequency voice traffic from the higher frequency internet data flowing on the line.

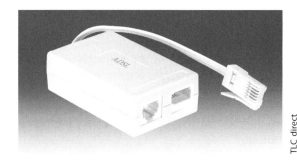

Figure 12.72 *Typical ADSL filter (this plugs into a phone socket, with two outputs, one for your ADSL modem and another for a telephone)*

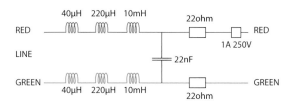

Figure 12.73 *Typical ADSL filter circuit*

Try this: Crossword

Across

1   Ratio between actual and apparent power. (5, 6)

5   A coil energises this component. (5)

7   Unit of illuminance. (3)

8   Type of heater. (7)

9   Solid-state switching device. (5)

10 Transformer winding connection. (5)

13 Two-terminal semiconductor device. (4)

14 Rectifier with four diodes. (6)

15 Substation transformer secondary winding connection. (4)

16 npn or pnp types. (10)

17 ac to dc conversion. (13)

Down

2   Type of trip. (8)

3   dc machine. (8)

4   ac motor rotor. (4)

6   Spur is a common type. (4)

8   Used to extend the range of an ammeter. (5)

9   Can be connected back to back. (9)

11 Incandescent lamp filament. (8)

12 Defect in an electrical circuit. (5)

13 Not alternating current. (6)

Congratulations you have now completed Chapter 12 which is the final chapter in this book. Correctly complete the self-assessment before continuing to the end test.

SELF ASSESSMENT

Circle the correct answers.

1   The value of the resistor in Figure 12.74 would be:

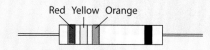

Red Yellow Orange

Figure 12.74 *Resistor*

a.   24Ω
b.   240Ω
c.   2 400Ω
d.   24kΩ.

2   The equivalent capacitance for the circuit shown in Figure 12.75 is:

a.   125μF
b.   200μF
c.   50μF
d.   100μF.

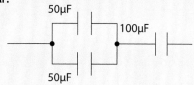

50μF

100μF

50μF

Figure 12.75 *Circuit*

3   One example of a silicon-controlled rectifier is a:

a.   diode
b.   full-wave bridge
c.   transistor
d.   thyristor.

4   The electronic component used in a dimmer switch to vary the light output of a lamp is a:

a.   thermistor
b.   triac
c.   zener diode
d.   fixed resistor.

5   One application for an inverter is:

a.   Ac to dc power supply
b.   half-wave rectifier circuit
c.   Dc to ac power supply
d.   motor temperature control.

End test

1. Transposing the formula $E = B\,I\,v$ to find I would result in:

   a. $I = \dfrac{EB}{v}$

   b. $I = \dfrac{E}{Bv}$

   c. $I = \dfrac{E-B}{v}$

   d. $I = \dfrac{Bv}{E}$

2. If the opposite side of a right-angled triangle is 80mm, and the adjacent side is 60mm, the hypotenuse side will be:

   a. 40mm

   b. 70mm

   c. 100mm

   d. 140mm

3. Simplifying $9^3 \times 9^2 \times 9$ gives

   a. 9^7

   b. 9^6

   c. 9^5

   d. 9^1

4. Inductance is measured in:

   a. farads

   b. henrys

   c. coulombs

   d. ohms

5. A multiplier is used to extend the range of a moving coil:

   a. ammeter

   b. voltmeter

   c. wattmeter

   d. ohmmeter

6. Which order of lever is a wheelbarrow an example of?

   a. first

   b. second

   c. third

   d. fourth

7. An electric motor has a mass of 180kg. What force is needed to lift it?

   a. 180N

   b. 176.58N

   c. 1 765.8N

   d. 1 800N

8. A 200V dc motor has an efficiency of 80% and draws a current of 10A. Its output power will be:

   a. 2 400W

   b. 2 000W

   c. 1 600W

   d. 1 200W

9. Which of the following statements is correct?

   a. neutrons are negatively charged

   b. electrons are neutral

   c. protons are positively charged

   d. protons are negatively charged

10. Which of the following cable insulation is most toxic when it burns?

☐ a. LSF

☐ b. LSHF

☐ c. PVC

☐ d. magnesium oxide

11. The lines of magnetic flux around a bar magnet:

☐ a. never form closed loops

☐ b. always cross one another

☐ c. have north to south direction

☐ d. are attracted to like poles

12. The average value of one complete ac cycle is:

☐ a. 0

☐ b. 0.637

☐ c. 0.707

☐ d. 1.732

13. Which of the following is not a renewable energy source for generating electricity?

☐ a. wave

☐ b. nuclear

☐ c. wind

☐ d. solar

14. In the figure below the impedance of the circuit is:

☐ a. 4.47Ω

☐ b. 14.14Ω

☐ c. 15.8Ω

☐ d. 20Ω

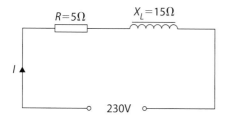

15. The power factor of an ac circuit is found by:

☐ a. $\sin\theta = \dfrac{kW}{kVA}$

☐ b. $\cos\theta = \dfrac{kVA_r}{kW}$

☐ c. $\cos\theta = \dfrac{kVA}{kW}$

☐ d. $\cos\theta = \dfrac{kW}{kVA}$

16. The supply voltage to a substation transformer is 11kV and has a line current of 100A to delta connected windings. The current flowing in the phase winding of the transformer is:

☐ a. 50A

☐ b. 57.8A

☐ c. 100A

☐ d. 173.2A

17. Balancing single-phase loads on a three-phase system is not carried out in order to:

☐ a. limit the current in the neutral conductor

☐ b. prevent excessive loading on any one line conductor

☐ c. reduce the voltage drop in the system

☐ d. increase the power loss in the system

18. The starting mechanism of a capacitor start split-phase motor consists of start windings in series with the:

☐ a. run windings, centrifugal switch and capacitor

☐ b. centrifugal switch and run windings

☐ c. centrifugal switch and capacitor

☐ d. run windings and capacitor

19. A three-phase induction motor can have the direction of rotation reversed by reversing:

☐ a. the start winding

☐ b. the run winding

☐ c. any two phases

☐ d. all three phases

20. The appropriate type of starter for starting a three-phase wound rotor motor against a heavy load is:

- ☐ a. direct-on-line
- ☐ b. rotor resistance
- ☐ c. star delta
- ☐ d. forward-reverse direct-on-line

21. The rated current of a fuse is the:

- ☐ a. maximum current causing the fuse to blow
- ☐ b. minimum current causing the fuse to blow
- ☐ c. maximum continuous current carrying capability of the fuse
- ☐ d. maximum current that can safely be interrupted by the fuse

22. Which type of fuse has the limitation that its fuse wire can deteriorate over a period of time?

- ☐ a. BS 88-3 cartridge
- ☐ b. BS 1362 cartridge
- ☐ c. BS 3036 semi-enclosed
- ☐ d. BS 88-2 HBC cartridge

23. Lumens per watt is the measurement of:

- ☐ a. luminance intensity
- ☐ b. luminous efficacy
- ☐ c. luminous flux
- ☐ d. luminance

24. A tungsten halogen lamp:

- ☐ a. has poor colour rendering
- ☐ b. runs at a high temperature
- ☐ c. is not suitable for display lighting
- ☐ d. requires expensive control equipment

25. Which of the following will not reduce stroboscopic effect?

- ☐ a. connecting fluorescent luminaries on one phase
- ☐ b. using lead/lag fluorescent luminaries
- ☐ c. tungsten lamps shining on any moving parts
- ☐ d. high-frequency fluorescent luminaries.

26. A 'loose wire' cable system is used for:

- ☐ a. immersion heaters
- ☐ b. convector heaters
- ☐ c. underfloor heating
- ☐ d. radiant heaters

27. Which of the following is correct for the bridge rectifier in the diagram?

	A	B	C	D
☐ a.	dc–	ac	dc+	ac
☐ b.	dc–	dc+	ac	ac
☐ c.	ac	ac	dc+	dc–
☐ d.	ac	dc+	ac	dc–

28. The device which is effectively two thyristors back to back is a:

- ☐ a. triac
- ☐ b. LED
- ☐ c. thermistor
- ☐ d. diode

29. The resistor shown could have a resistance between:

- ☐ a. 2 156 and 2 244Ω
- ☐ b. 2 090 and 2 130Ω
- ☐ c. 1 980 and 2 420Ω
- ☐ d. 1 760 and 2 640Ω

30. A thyristor will conduct more current when triggered at a firing angle of:

- ☐ a. 5°
- ☐ b. 45°
- ☐ c. 90°
- ☐ d. 120°

Formulae/Equations used in book 8

Chapter 1

Ohm's law:
$$I = \frac{V}{R} \therefore V = I \times R \text{ or } R = \frac{V}{I}$$

Formula relating power to torque:
$$P = \frac{2\pi NT}{60} \therefore T = \frac{60P}{2\pi N}$$

Impedance:
$$Z = \sqrt{R^2 + X_L{}^2} \therefore X_L = \sqrt{Z^2 - R^2} \text{ or } R = \sqrt{Z^2 - X_L{}^2}$$

Pythagoras' theorem:
$$c = \sqrt{a^2 + b^2} \therefore a = \sqrt{c^2 - b^2} \text{ or } b = \sqrt{c^2 - a^2}$$

Trigonometric ratios:
$$\sin \theta = \frac{\text{opposite}}{\text{hypotenuse}}$$
$$\cos \theta = \frac{\text{adjacent}}{\text{hypotenuse}}$$
$$\tan \theta = \frac{\text{opposite}}{\text{adjacent}}$$

Chapter 2

Conversion of temperature units:
$$\text{Celsius } (°C) = 5/9 \, (°F - 32) \text{ and}$$
$$\text{Fahrenheit } (°F) = 9/5 \times °C + 32$$

Area:
$$\text{Area of a rectangle} = \text{length} \times \text{breadth}$$

Area of a circle:
$$\text{Area} = \pi r^2 \text{ or } \frac{\pi d^2}{4} \text{ (also CSA of a circle)}$$

Area of a triangle:
$$\text{Area} = \frac{1}{2} \text{ base} \times \text{height}$$

Volume of a cuboid:
$$\text{Volume} = \text{length} \times \text{breadth} \times \text{depth (or height)}$$

Volume of a cylinder:
$$\text{Volume} = \frac{\pi d^2 \times l}{4} \text{ or } \pi r^2 \times l$$

Resistance:
$$\text{Resistance} = \frac{\text{resistivity} \times \text{length } (l)}{\text{cross} - \text{sectional area } (A)} \text{ or } R = \frac{\rho l}{A}$$

Resistance after a temperature rise:
$$\frac{R_1}{R_2} = \frac{1 + a\,t_1}{1 + a\,t_2}$$

Quantity of electricity:
$$\text{amperes} = \frac{\text{quantity of electricity in coulombs}}{\text{time in seconds}}$$
$$\text{or } I = \frac{Q}{t} \text{ or } Q = It$$

Power:
$$\text{Power} = \frac{\text{energy}}{\text{time}} \therefore \text{Energy} = \text{power} \times \text{time}$$

Frequency:
$$f = \frac{1}{T} \therefore \text{Periodic time } T = \frac{1}{f}$$

Inductive reactance:
$$X_L = 2\pi f L$$

Charge on a capacitor:
$$Q = CV \therefore C = \frac{Q}{V}$$

Capacitive reactance:

$$X_C = \frac{1}{2\pi f C} \text{ when capacitance is in Farads (F)}$$

$$X_C = \frac{10^6}{2\pi f C} \text{ when capacitance is in microfarads } (\mu F)$$

Impedance (as in Chapter 1):

$$Z = \sqrt{R^2 + X_L^2} \text{ (with just resistance (R) and inductance}$$
$$\text{(L) in the circuit)}$$

$$Z = \sqrt{R^2 + X_C^2} \text{ (with just resistance (R) and capacitance}$$
$$\text{(C) in the circuit)}$$

$$Z = \sqrt{R^2 + (X_L - X_C)^2} \text{ (R, L and C circuit with } X_L$$
$$\text{greater than } X_C\text{)}$$

$$Z = \sqrt{R^2 + (X_C - X_L)^2} \text{ (R, L and C circuit with } X_C$$
$$\text{greater than } X_L\text{)}$$

Actual or true power of a dc circuit:

$$P = V \times I \therefore I = \frac{P}{V} \text{ or } V = \frac{P}{I}$$

Power factor:

$$PF = \cos\theta = \frac{kW}{kVA} \text{ or } \cos\theta = \frac{W}{VA} \text{ or } \cos\theta = \frac{R}{Z}$$

Power in ac circuit:

$$kVA = \sqrt{kW^2 + kVA_r^2}$$

Impedance:

$$Z = \frac{V}{I} \therefore V = I \times Z \text{ or } I = \frac{V}{Z}$$

Chapter 3

Force:

$$F = m \times a$$
$$\text{moment} = \text{force} \times \text{distance}$$
$$= f_1 d_1 \text{ or } f_2 d_2$$

Weight:

$$Weight = mass \times 9.81\,N$$

Work done:

$$Work\ done = force\ applied \times distance\ moved$$
$$W = F \times d$$

Energy:

$$Energy = Power \times time \therefore Power = \frac{Energy}{time}$$
$$or\ Joules = Watts \times time,\ J = W \times t \therefore W = \frac{J}{t}$$

Kinetic energy:

$$KE = \frac{1}{2}mv^2$$

Gravitational potential energy:

$$GPE = mgh$$

Efficiency:

$$\text{Efficiency } (\eta) = \frac{\text{useful energy output}}{\text{total energy input}} \times 100\%$$

$$P \text{ (in watts)} = \text{Time rate of doing work}$$
$$= \frac{\text{work done (in joules)}}{\text{time taken (in seconds)}}$$
$$or\ watts = \frac{\text{joules}}{\text{time}}$$
$$W = \frac{J}{t}$$
$$\text{efficiency } (\eta) = \frac{\text{output power}}{\text{input power}} \times 100\%$$

Chapter 4

Resistance and resistivity of electrical circuit:

$$R_{circuit} = R_1 + R_n + R_{Load}$$

Total resistance in a series circuit:

$$R_T = R_1 + R_2 + R_3, \text{ etc}$$

Total resistance in a parallel circuit (product over sum method for two resistors only):

$$R_T = \frac{R_1 \times R_2}{R_1 + R_2}$$

Total resistance in a parallel circuit (more than two resistors):

$$\frac{1}{R_T} = \frac{1}{R_1} + \frac{1}{R_2} + \frac{1}{R_3} + \frac{1}{R_4} \ldots \text{etc}$$

Total current in a series circuit:

$I_T = I_1 = I_2 = I_3$ (all of the currents I are the same value as the supply current)

Total current in a parallel circuit:

$I_T = I_1 + I_2 + I_3$ (sum of currents = supply current)

Power in dc circuits:

$$P = I^2 R$$

$$\text{or } P = VI$$

$$\text{or } P = \frac{V^2}{R}$$

Chapter 5

Magnetic flux density:

$$B = \frac{\Phi}{A}$$

Force on a current-carrying conductor in a magnetic field:

$$F = BII$$

Induced emf:

$$E = Blv$$

Induced emf by rate of change of magnetic flux:

$$E = \frac{N(\Phi_2 - \Phi_1)}{t}$$

Average and RMS values:

$$V_{av} = V_{max} \times 0.637 \therefore V_{max} = \frac{V_{av}}{0.637}$$

$$V_{r.m.s} = V_{max} \times 0.707 \therefore V_{max} = \frac{V_{rms}}{0.707}$$

$$I_{av} = I_{max} \times 0.637 \therefore I_{max} = \frac{I_{av}}{0.637}$$

$$I_{r.m.s} = I_{max} \times 0.707 \therefore I_{max} = \frac{I_{rms}}{0.707}$$

Chapter 6

Basic transformer ratio equation:

$$\frac{V_P}{V_S} = \frac{N_P}{N_S} = \frac{I_S}{I_P}$$

$$\frac{V_P}{V_S} = \frac{N_P}{N_S} \therefore V_S = V_P \times \frac{N_S}{N_P}$$

$$\frac{V_P}{V_S} = \frac{I_S}{I_P} \therefore I_S = \frac{V_P}{V_S} \times I_P$$

Volts per turn:

$$\frac{V_P}{N_P} = \frac{V_S}{N_S}$$

kVA rating of a single-phase transformer:

$$kVA = \frac{V \times I}{1000}$$

Percentage efficiency:

$$\% \text{ efficiency} = \frac{\text{Power out}}{\text{Power in}} \times 100$$

Transformer efficiency including losses:

$$\text{Efficiency} = \frac{\text{Output}}{\text{Output} + \text{losses}} \times 100\%$$

Chapter 7

Reactance:

$$X_L = 2\pi fL \text{ and so } L = \frac{X_L}{2\pi f}$$

$$X_C = \frac{1}{2\pi fC} \text{ and so } C = \frac{1}{2\pi fX_C} \text{ or } \frac{10^6}{2\pi fX_C} \text{ when C is in } \mu F$$

Impedance:

$$Z = \frac{V_r}{I}, \; X_L = \frac{V_L}{I} \text{ and } R = \frac{V_R}{I}$$

$$\therefore V_T = IZ, \; V_L = IX_L \text{ and } V_R = IR$$

$$\therefore I = \frac{V_r}{Z}, \; I = \frac{V_L}{X_L} \text{ and } I = \frac{V_R}{R}$$

Refer to Chapter 2 for other formulae.

Single-phase ac power:

$$P = VI \cos \theta \therefore I = \frac{P}{V \times \cos \theta}$$

Purely resistive single-phase ac circuit:

$$P = VI \therefore I = \frac{P}{V}$$

Delta and star, voltage and current relationships:

For delta : $\quad V_L = V_P$

$$I_L = \sqrt{3} \times I_P \quad \text{or } I_P = \frac{I_L}{\sqrt{3}}$$

For star : $\quad I_L = I_P$

$$V_L = \sqrt{3} \times V_P \quad \text{or } V_P = \frac{V_L}{\sqrt{3}}$$

Three-phase power, for star or delta balanced systems:

$$\text{3-phase power} = \sqrt{3} V_L I_L \cos \theta \therefore I_L = \frac{3 \text{ Phase power}}{\sqrt{3} \, V_L \cos \theta}$$

Chapter 8

Fractional (per unit) slip:

$$S = \frac{N_s - N_r}{N_s}$$

Percentage slip:

$$S(\%) = \frac{N_s - N_r}{N_s} \times 100$$

Synchronous speed:

$$N_s = \frac{f}{P} \times 60 \quad \text{in rev/min}$$

$$n_s = \frac{f}{P} \quad \text{in rev/s}$$

Rotor speed:

$$n_r = n_s(1 - s)$$

Chapter 9

$$\text{Fusing factor} = \frac{\text{fusing current}}{\text{current rating}}$$

$$\therefore \text{ Fusing current} = \text{fusing factor x current rating}$$

Chapter 10

Luminous efficacy:

$$\text{Efficacy (1 m/W)} = \frac{\text{light output (lm)}}{\text{electrical input (W)}}$$

Inverse square law:

$$E = \frac{I}{d^2}$$

Cosine law:

$$E = \frac{I}{d^2} \times \cos \theta$$

Lumen method (to find number of lamps required):

$$N = \frac{E \times A}{F \times UF \times MF}$$

Chapter 12

Resistors in series and parallel covered in Chapter 4

Capacitors in parallel:

$$C_{Total} = C_1 + C_2 + C_3 + C_4..... \text{ etc.}$$

Capacitors in series:

$$\frac{1}{C_{Total}} = \frac{1}{C_1} + \frac{1}{C_2} + \frac{1}{C_3} + \frac{1}{C_4}...\text{etc.}$$

Two capacitors in series (product over sum method):

$$C_T = \frac{C_1 \times C_2}{C_1, +, C_2}$$

Answer section

Chapter 1

Recap Page 2

fraction
numerator   denominator
simplifying
integer
20
$\frac{1}{5}$ 0.2

Try this Page 4

72

Try this Page 4

1 $\frac{11}{15}$

2 $\frac{13}{30}$

3 $\frac{3}{8}$

Try this Page 5

1 $\frac{21}{44}$

2 $\frac{15}{56}$

3 4

4 $\frac{8}{27}$

Try this Page 6

1 $16\frac{3}{10}$

2 $3\frac{3}{8}$

3 $2\frac{2}{3}$

Try this Page 6

1 12

2 $2\frac{2}{3}$

Try this Page 6

1 6.143

2 11.167

Try this Page 7

1 $\frac{7}{8}$

2 $\frac{1}{8}$

1 62.5%

2 68.75%

Try this Page 9

1 $Y = \dfrac{BD}{M}$

2 $Y = X - A + B$

3 $Y = \dfrac{WX}{ABC}$

4 $Y = \dfrac{F(A + B)}{2E}$

Try this Page 9

1. $x = 9$
2. $x = 6$
3. $x = 3$

Try this Page 11

$$F_2 = F_1 - \left(\frac{60P}{2\Pi rN}\right)$$

Try this Page 13

1. $IR + Ir$
2. $5y - 27$
3. $2p + 14q$

Try this Page 15

1. a) 17
   b) 394
2. a) 729
   b) 29 791
3. a) 7^7
   b) 9^6
4. a) $\dfrac{1}{10^{-4}}$

   b) 10^{-7}

Try this Page 15

1. 9.78×10^8
2. 3.45×10^{-6}
3. 1×10^{-3}

Try this Page 16

1. 12cm
2. 8cm

Try this Page 18

1. a) 0.866
   b) 13.856cm
2. a) 1.6
   b) 9.6cm

SELF ASSESSMENT Page 19

1. c. $\dfrac{1}{100000}$
2. b. 4
3. b. 5^2
4. a. $R = \dfrac{P}{I^2}$
5. d. 250V

Chapter 2

Recap Pages 20 and 21

1. 720
2. $e = d - a + b$
3. $\dfrac{1}{100000}$
4. a) 1.2
   b) 4.8 cm

Try this Page 22

50°F

Try this Page 24

5m/s^2

Try this Page 24

Milliampere
Gigawatt
Kilovolt
Picofarad

Try this Page 25

1. 1875mm^2
2. 9.62mm^2

Try this Page 26

1. 9m^3
2. 0.29m^3

Try this Page 27

Unit	Unit symbol	Variable symbol
kilogram	kg	m
square metre	m^2	A
metre/second	m/s	v
kelvin	K	t
kilogram/cubic metre	kg/m^3	ρ

Try this Page 28

4mm

Try this Page 29

0.029Ω (29mΩ)
88Ω

Try this Page 30

1   0.071Ω
2   15 000C

Try this Page 31

100Hz

Try this Page 32

314.2Ω

Try this Page 33

18.4mC

Try this Page 34

1   144.67Ω
2   21.22Ω

Try this Page 35

223.6Ω

Try this Page 36

10.97Ω

Try this Page 37

0.9 leading

Try this Page 38

0.9 lagging

Try this Page 38

1   21.8 VA$_r$
2   0.9 leading

Try this Page 39

51kW

Try this Page 41

a) R = 5Ω & b) Z = 13Ω.

Try this Page 43

a)   moving coil
b)   moving iron
     moving coil
     moving iron
     moving coil
     digital   power
     ammeter
     extend   voltmeter

Try this Page 45

0.68 lagging
GS38
current   disconnect   circuit
actual   apparent
power   time   kWh

SELF ASSESSMENT Page 45

1   c. metres per second
2   d. 5625mm^2.
3   c. 8m
4   c. both statements are correct
5   b. actual power

Chapter 3

Recap Page 46

capacitance
frequency
resistivity
ohm
joule
watt
apparent power
resistance
power
frequency

Try this Page 48

1   0.1m/s^2
2   784.8N
3   3433.5N

Try this Page 49

1   approx 6 N
2   approx 6 N

the quantity of matter that a body contains and it is measured in kilograms.
an influence tending to cause the motion of a body.

the product of the mass of an object and the force of the earth's gravitational force.

direction

Try this Page 51

136.25N

Try this Page 52

1   30kg
2   9kg
3   4-pulley system

Try this Page 55

3924N

Try this Page 55

147.15kJ

Try this Page 56

1   1471.5N
2   4414.5J

Try this Page 57

18MJ

Try this Page 58

1   137 700J
2   2754N

Try this Page 58

60J

Try this Page 60

4087.5W

Try this Page 61

12262.5W

Try this: Crossword Page 61

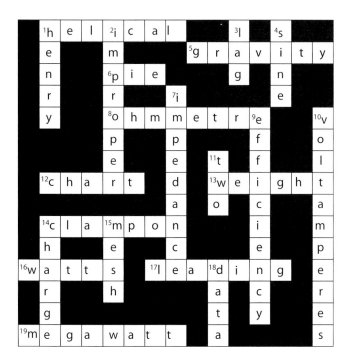

SELF ASSESSMENT Page 62

1   c. 1962.00N
2   a. 10kg
3   b. 20.25J
4   b. 73.58W
5   d. one fixed and one movable pulley

Chapter 4

Recap Page 63

1   matter   kilograms
2   1471.5N

3   work done = 0.02J, power = 0.0044W
    (4.4mW)
4   slipping   exact   ratios

Try this Page 65

protons   neutrons   electrons
Protons   neutrons   electrons

Try this Page 65

pass   copper
insulators
resistance   conductors

Task Page 69

1   **300/500V,**
2   **600/1000V,**
3   **300/500V,**
4   **500V,**
5   **750V.**

Try this Page 71

low   high   copper   aluminium
sheathed   protection   smoke   gases
smoke   fume   smoke   halogen
magnesium   moisture   atmosphere
fibre   185°C
optic   current   light

Try this Page 72

20.688Ω

Try this Page 72

1   9.2A
2   5A

Try this Page 73

1.			24Ω
2.			3Ω
3.	88V		
4.		1.44A	
5.			48Ω

Try this Page 74

$R_1 = 0.5Ω$, $R_2 = 2Ω$

Try this Page 75

1   20Ω
2   6A
3   $I_1 = 2A$ and $I_2 = 4A$

Try this Page 76

1   10A, 6A, 4A
2   20A
3   2.4Ω

Try this Page 77

3.6Ω

Try this Page 78

I = V/R
adding
same   voltage
current

$P = I^2R$ $P = \dfrac{V^2}{R}$

Try this Page 79

1   2.4V
2   9.6V

Try this Page 80

thermal, chemical, magnetic
electron   electrolyte   negative   positive

SELF ASSESSMENT Page 81

1   d. 29
2   c. Magnesium oxide
3   b. 2.5A
4   a. 25W
5   a. 22.8V

Chapter 5

Recap Page 82

1   positively   negatively   neutral
2   smoke free
3   current   voltage   voltage   current
4   20V

Try this Page 85

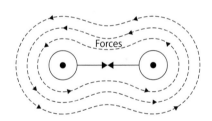

Try this Page 85

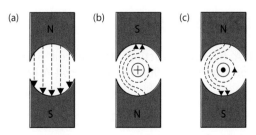

Try this Page 86

attract   repel
field   current
fingers   thumb   field

Try this Page 86

0.1T

Try this Page 87

Φ   field   weber
B   tesla
increasing   more

Try this Page 88

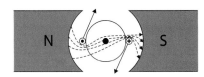

Try this Page 89

312.5A

Try this Page 90

6V

Try this Page 90

30V

Try this Page 91

In coil A by self induction, and in coil B by mutual induction due to the flux set up by coil A linking with coil B.

Try this Page 91

E = Blv
self   mutual
henry   A/s   1V

Try this Page 92

1   To connect the coil to the meter
2   360°
3   Coil A away from you, coil B towards you

Try this Page 94

maximum = 19.8A and average = 12.6A

SELF ASSESSMENT Page 95

1   c. tesla
2   b. slip rings
3   c. motion direction
4   a. 282V
5   a. 16A

Chapter 6

Recap Page 96

1   field   current   emf   motion
2   motion   magnetic
    coil   poles
    magnetic   windings
3   RMS   average
4   the ac RMS current equals the dc current value and gives the same heating effect.

Try this Page 98

alternator
Fossil   coal   oil
cheap

Try this Page 100

25kV
132kV   400kV   275kV

Try this Page 101

1   253V
2   216.2V

Try this: Wordsearch Page 103

A	H	Y	D	R	O	E	L	E	T	R	I	C	H	S
S	E	O	D	I	S	T	R	I	B	U	T	I	O	N
U	E	G	N	F	G	I	N	E	N	F	S	T	S	P
B	N	L	S	B	S	N	U	C	L	E	A	R	U	O
S	P	O	E	T	E	V	O	S	N	T	B	B	P	W
T	G	H	I	U	O	S	I	A	L	W	A	C	E	E
A	T	I	S	T	F	S	H	E	P	L	N	G	R	R
T	D	T	P	I	A	L	D	O	I	N	S	I	G	S
I	T	I	N	H	S	R	I	S	H	S	B	C	R	T
O	F	O	U	G	I	H	E	S	I	O	N	D	I	A
N	U	N	D	D	S	A	D	N	S	T	F	D	D	T
F	L	D	P	L	H	S	N	T	E	O	S	I	P	I
I	N	D	U	S	T	R	I	A	L	G	F	E	S	O
N	T	S	I	V	S	T	A	R	P	O	I	N	T	N
T	R	A	N	S	M	I	S	S	I	O	N	P	I	H

Try this Page 107

three   twelve
chemical
negative   positive   electrolyte

Try this Page 109

1   4:1
2   2000 turns

Try this Page 110

1   57.5V
2   0.46

Try this Page 110

air   tank   cooling   circulates
alternating   primary   flux   core   secondary   emf
mutual
-down   primary   -up   secondary
volts   secondary

Try this Page 111

Actual value is 13.8kVA, therefore 15kVA

Try this Page 113

1   400V
2   230.9V

Try this Page 114

95.65%

Try this Page 114

15kW, 97%
isolating   mains   earth
power out
winding

$$\text{Efficiency} = \frac{\text{Output}}{\text{Output} + \text{Losses}} \times 100\,\%$$

Try this: Crossword Page 117

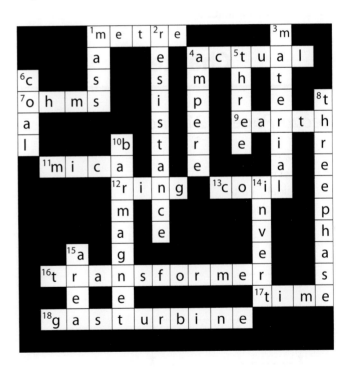

SELF ASSESSMENT *Page 118*

1   c. 25kV
2   c. x series and y parallel
3   b. 62.5V
4   d. hydro-electric
5   b. 97.4%

Progress Check Pages 119 and 120

1 d. $11\dfrac{3}{8}$
2 b. 3
3 b. 26.57°
4 a. kilograms
5 c. 199.8Ω
6 d. apparent power
7 d. 68.5Ω
8 b. 20 000J
9 c. 12kW
10 c. ductile
11 b. 6Ω
12 b. 200W
13 a. field
14 b. upwards
15 c. 297V
16 d. earth and neutral
17 a. 400kV
18 d. 50V
19 c. 78.48N
20 b. it has an impact on the landscape

Chapter 7

Recap Pages 121 and 122

1 delta   star
2 protective
   terminal   earth
   earth   electrode
   neutral
   winding
   line

3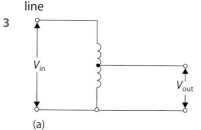
(a)

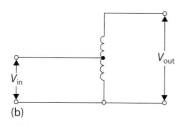
(b)

4 19.2 kVA

Try this Page 123

resistance   reactance   capacitive   current   opposition   impedance   ohms (Ω)

Try this Page 123

	628 Ω			0.366 A		
		318 Ω		0.723 A		
			240V		0.31H	
		498 Ω				8 μF
30 Ω				4.4 A		
		50 Ω	24V			

Try this Page 125

80V

Try this Page 128

1 29.5 Ω; 2 7.8 A; 3 (i) 122.5V, (ii) 195V
series   same
parallel   voltage
leads   lags

Try this Page 129

1 a) 0.6 lagging
   b) 4kVA$_r$
2 a) 3.75kVA
   b) 2.25kVA$_r$

Try this Page 131

1 $I_C = I_L = 20A$, 2 = 276.75μF

Try this Page 132

16.3A
capacitor   supply
larger   heavier
$P = VI \cos \theta$ $I = \dfrac{P}{V \times \cos\theta}$

Try this Page 133

1   a) 138.73V
    b) 245.7V
    c) 173.4V
2   a) 718V
    b) 398V
    c) 190.3V

Try this Page 134 and 135

1   a) 57.8A
    b) 138.7A
    c) 1156A
2   a) 86.5A
    b) 415.2A
    c) 1.73kA

Try this Page 136

1   14.4kW
2   43.2kW

Try this Page 137

1   57.73A
2   33.33A
3   39.995kW
    50.35A   29A.

Try this Page 139

Approx. 13A (2.6 cm)
star   delta
$P = \sqrt{3}\, V_L I_L \cos \theta$
limit   neutral
load
current   zero

SELF ASSESSMENT Page 140

1   d. 15.8Ω
2   b. $P = VI \cos \theta$
3   c. 0.8
4   c. 10kVA
5   a. 0A

Chapter 8

Recap Pages 141 and 142

1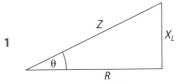

2   a = 100V, b = 207V
3   (a) I_L = 86.6A, I_P = 50A; (b) 60kW

4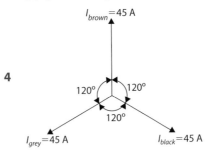

Try this Page 144

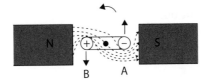

Try this Page 145

electrical   mechanical   mechanical   electrical
rotating   changes   current   armature
alternating   armature   direct

Try this Page 147

weaker   increases   series   shunt

Try this Page 148

separately external power provide magnetic flux
self voltage residual field

Try this Page 150

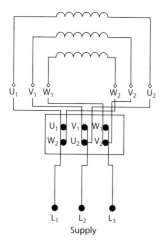

Try this Page 151

rotation two
rotating stator
interacting rotor
rotor same magnetic field

Try this Page 154

winding terminal star delta
cage- high low
double lower higher single
wound all external cut out
synchronous self power factor

Try this Page 155

25 rev/sec

Try this Page 156

4%

Try this Page 156

1   50 rev/s
2   48.5 rev/s.
synchronous frequency number pairs poles

Try this Page 159

two start run parallel
centrifugal switch burning
either start run
Universal high

Try this Page 161

light
overloads excessive
No-volt automatic starting failure
remote parallel series.

Try this Page 163

starting current starting stator star full
speed stator delta

Try this Page 164

reduced star/delta
stator star voltage winding full delta line
400V winding
resistance wound heavy loads
current first smooth torque load

SELF ASSESSMENT Page 166

1   b. compound wound
2   c. high starting current and a low starting torque
3   b. 58%
4   c. 1000 rev/min
5   d. poor speed control

Chapter 9

Recap Page 167

1   Change armature or field windings but not both.
2   To connect the rotor windings to the external resistances.
3   Starts in parallel, stops in series.
4   8.33 rev/s

Try this Page 170

operated   opens   contacts   current   circuit   coil
magnetic   armature   armature   close   circuit
electrically   current
wire   current   solenoid

Try this Page 175

persons   property   enclosed   cartridge   high breaking
capacity
maximum   interrupted
overload   short circuit

Try this Page 177

residual current device   line   neutral   fault   earth
RCD   MCB

Try this: Crossword Page 178

SELF ASSESSMENT Page 179

1   d. triple-pole contactor.
2   c. moving armature
3   b. a ceramic body
4   b. silica sand
5   b. the line and neutral currents are not equal

Chapter 10

Recap Page 180

1   coil   magnetic   armature   pole   open   close
2   cheap   repair
3   high short circuit   high
4   thermal   overload   magnetic
5   overload   short circuit   earth

Try this Page 182

62.5 lm/W

Try this Page 183

lumens   lux
luminous efficacy
discharge   ultraviolet   visible

Try this Page 184

(a) 133 lux
(b) 48 lux

Try this Page 185

P_1 = 133.33 Lux, P_2 = 28.8 Lux

Try this Page 186

39.77 so 40 lamps required

Try this Page 187

inverse square

$$E = \frac{I}{d^2} \times \cos\theta$$

luminaire   surfaces
number   lamsps

Try this Page 190

gas   evaporation   arcing
high   combustible
gases   light
inductor   discharge   limit   lamp   struck

Try this Page 194

fluorescent.
starter interference capacitor factor
output supply temperature
decreases pressure arc

Try this Page 195

yellow
poor high sodium
discharge neon argon vaporizes heat

Try this Page 198

80% tungsten magnetic electronic
electronic flicker
energy compact
ac/dc

SELF ASSESSMENT Page 198

1   c. efficacy is measured in candela/m^2
2   b. low pressure sodium
3   d. 200 lux
4   b. 80%
5   d. 6400K cool white LED lamp.

Try this Page 203

radiant tungsten heat quartz
warm circulating
bricks thermal
cable mats underfloor

Try this Page 206

immersion time hot
bimetallic two expansion

Try this Page 208

cylinder temperature hot water
pump boiler

SELF ASSESSMENT Page 209

1   c. storage heater
2   b. rod
3   d. laminate flooring
4   d. motorized valve.
5   b. convection

Chapter 11

Recap Page 199

1    $P_1 = 62.5$ lux, $P_2 = 15.26$ lux
2   type height size colour texture ceilings size
3   capacitor power correction starts discharge
     limits lamp lit filaments choke discharge lamp
4   Answers can include any three from: traffic signals,
     railway signals, street lighting, floodlights, stage lights,
     exit signs, information signs, pedestrian crossing and
     speed limit signs.

Try this Page 203

$$R = \frac{V^2}{P} = \frac{230^2}{320} = 165\Omega$$

Chapter 12

Recap Page 210

1   a) radiation   b) convection   c) conduction
2   cool bottom heated elements rises cools
     circulates reheats
3   a) takes a long time to re-heat   b) there is heat loss
     even when not in use
4   To control the temperature of domestic hot water

Try this Page 212

1   Actual value = 0.21W, 0.25W standard resistor
     suitable
2   Actual value = 3.7W, 5W high power resistor suitable

Try this Page 214

voltage   current
control   series
twice   half
temperature   semiconductor

Try this Page 215

51 000Ω or 51kΩ
2400Ω or 2.4kΩ
27Ω

Try this Page 216

1 44650Ω   49350Ω
2 24.3Ω   29.7Ω

Try this Page 217

1   10Ω ± 5%
2   1.5kΩ ± 10%
3   1MΩ ± 20%

1   1M5J
2   R47G
3   10MF

Try this Page 220

4.8µF

Try this Page 223

four   diodes   ac   dc   one
output
smoother   stabilizes

Try this Page 225

flat   shorter
a) emitter   b) base   c) collector.

Try this Page 228

1   positive half cycle
2   both positive and negative half cycle

Try this: Crossword Page 231

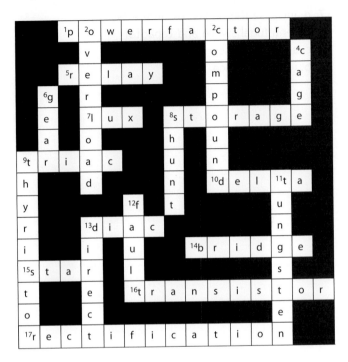

SELF ASSESSMENT Page 232

1   d. 24kΩ
2   c. 50µF
3   d. thyristor
4   b. triac
5   c. Dc to ac power supply

End test

1   b. $I = \dfrac{E}{Bv}$
2   c. 100mm
3   b. 9^6
4   b. henrys
5   b. voltmeter
6   b. second
7   c. 1765.8N
8   c. 1600W

9   c. protons are positively charged

10 c. PVC

11 c. have north to south direction

12 a. 0

13 b. nuclear

14 c. 15.8Ω

15 d. $\cos\theta = \dfrac{kW}{kVA}$

16 b. 57.8A

17 d. increase the power loss in the system

18 c. centrifugal switch and capacitor

19 c. any two phases

20 b. rotor resistance

21 c. maximum continuous current carrying capability of the fuse

22 c. BS 3036 semi-enclosed

23 b. luminous efficacy

24 b. runs at a high temperature

25 a. connecting fluorescent luminaires on one phase

26 c. underfloor heating

27 a. dc–   ac   dc+   ac

28 a. triac

29 a. 2156 and 2244Ω

30 a. 5°

Glossary

actual power the power dissipated in the circuit resistance, also called true power measured in watts *(W)* or kilowatts *(kW)*

algebra a branch of mathematics which employs the use of symbols or letters to solve mathematical problems

analogue instrument an instrument that takes an electric current and uses it to produce a mechanical deflection of a pointer across a calibrated scale

apparent power the combined effects of the resistive and reactive power in an ac circuit, measured in volt-amperes *(VA)* or kilovolt-amperes *(kVA)*

armature rotating part of a dc machine or universal motor; also the moving part of a relay that makes or breaks the relay contacts

auto-transformer a transformer which has only one winding, capable of stepping up or stepping down a voltage

average value is 0.637 of an ac voltage or current taken over half a cycle, and when taken over one full cycle is zero (since + 0.637 − 0.637 = 0)

battery a single cell or a combination of two or more cells connected to produce electrical energy

breaking capacity the maximum current that can safely be interrupted by a fuse

cage rotor the revolving part of a single or three phase ac induction motor

capacitance *(C)* a measure of a capacitor's ability to store an electric charge; its unit is the farad *(F)*

capacitive reactance *(X_C)* the opposition (offered by a capacitor) to the changing current in an ac circuit, measured in ohms *(Ω)*

capacitor an electronic component that stores electrical charge.

capacitor start/induction run motor a single-phase ac motor with two stator windings, a run winding and a start winding with a series capacitor

cartridge fuse a the fuse with fuse wire enclosed in a ceramic or glass body

centrifugal switch a switch that cuts out the starting circuit of a single-phase ac induction motor

choke an inductor used in discharge lighting circuits, and is basically a coil of wire wound round an iron core

circuit breaker (cb) an overcurrent protection device with a magnetic and a thermal trip

colour rendering the ability of a lamp or light source to show the true colours of objects

compound dc machine a dc motor or generator with field windings connected in series and parallel with the armature

conductor the part of an electrical cable that carries the current

contactor an electrically operated switch which is used for switching (high current) power circuits

convector heater a space heater from which heat is transferred to the surrounding air by convection

current *(I)* a measure of the rate of flow of electrons through a conductor, measured in amperes *(A)*

delta three-phase winding connection where the voltage across each phase winding V_P is the same as the voltage between the supply lines V_L

density *(ρ)* a measure of the compactness of a substance, measured in kg/m^3

diac a two terminal bidirectional diode, often used for triggering a triac

digital instrument an instrument used to measure electrical variables and display the measured readings usually on a liquid crystal display (LCD)

diode a semiconductor device that permits current to flow through it in one direction only (anode to cathode)

direct-on-line (DOL) starter a starter which connects the 400V three-phase supply lines directly to the motor windings of a three-phase ac induction motor on starting

discharge lamp a lamp in which light is produced by an electrical discharge in a gas-filled glass envelope

distribution system a system that provides three-phase and single-phase supplies to consumers

efficiency *(η)* the ratio between output power and input power

electromagnet a coil of wire usually wound around an iron or steel core

electrons negatively charged particles of atoms

electromotive force (emf) the force that moves electrons, measured in volts *(V)*

energy *(W)* the ability to do work, measured in joules *(J)* or more conveniently in kilowatt-hours *(kWh)*

fluorescent luminary/lamp a low-pressure mercury vapour discharge lamp

force *(F)* an influence tending to cause the motion of a body, measured in newtons *(N)*

fossil fuel a fuel that is the remains of some living organisms that lived millions of years ago (examples are coal, oil and natural gas)

fraction part of a whole number and is made up of two numbers, the numerator on the top (upstairs) and the denominator on the bottom (downstairs)

frequency *(f)* the number of cycles per second *(cps)* *(50 cps = 50 Hertz)*

fuse rating the maximum continuous current carrying capability of a fuse, without it blowing

fusing current the minimum current causing the fuse to blow

gear a rotating machine part with cut teeth, or cogs, which mesh with another toothed part in order to transmit torque (or rotational motion)

generator a rotating machine that converts mechanical energy to electrical energy

hydroelectric electricity generated by the pressure of falling water usually from a reservoir high up in the mountains

illuminance (E) a measure of the amount of light falling from a light source (luminaire) onto a surface, measured in lux (lx)

immersion heater an electric heater used to heat the domestic hot water in the storage cylinder

impedance (Z) the effective opposition to alternating current flow of all the components (resistance, inductance and capacitance) in a circuit, measured in ohms (Ω)

incandescent lamp a lamp with a fine filament wire that heats up until it is white hot and gives off light

indices a 'shorthand' way of writing very large and very small numbers

inductance (L) the property of a circuit by which a change in current induces an emf into the circuit (self inductance) or into a neighbouring circuit (mutual inductance); its unit is the henry (H)

induction start and run motor a single-phase ac induction motor with a start and run winding

inductive reactance (X_L) the opposition (offered by an inductor) to the changing current in an ac circuit, measured in ohms (Ω)

inductor a coil of wire with an iron, ferrite or air core

inverter electronic component that converts a direct current into an alternating current

kinetic energy energy of an object due to its motion

lever a bar pivoted so as to be able to rotate about a point

light emitting diode (LED) a PN junction diode which gives off light when forward biased

lumen method used to determine the number of lamps that should be installed for a given room or area

magnetic flux (Φ) a measure of the magnetic field produced by a permanent magnet or electromagnet, measured in webers (Wb)

magnetic flux density (B) the strength of a magnetic field is measured in terms of its magnetic flux density, measured in teslas (T)

mass (m) the quantity of matter that a body contains, measured in kilograms (kg)

motor a rotating machine that converts electrical energy to mechanical energy

multiplier a series resistor used to extend the range of a moving coil voltmeter to measure voltages higher than the instrument's movement is designed for

percentage a way of expressing a number as a fraction of 100 (the denominator is always 100)

potential energy stored energy

power (P) the rate at which work is done in watts (W), kilowatts (kW) or megawatts (MW)

power factor the ratio between actual or true power and apparent power

pulley a mechanism composed of a wheel on an axle or shaft that usually has a grooved rim around its circumference

Pythagoras' theorem this theorem states that in any right-angled triangle, the square of the hypotenuse is equal to the sum of the squares of the other two sides

RCBO the residual current circuit breaker with overload protection (RCBO) combines both RCD protection and CB protection in a single unit

reactive power the wattless power in an ac circuit measured in reactive volt-amperes (VAr) or reactive kilovolt-amperes (kVAr)

rectifier an electronic device containing one or more diodes which is used to convert an alternating current to a direct current

relay an electrically operated switch that opens and/or closes contacts to control the flow of current in one or more separate circuits

residual current device (RCD) a device which is able to detect very small earth fault currents and automatically disconnect the supply

resistance (R) the opposition to a current in a circuit, measured in ohms (Ω)

resistivity (ρ) the ability of a material to resist the passage of an electric current, measured in ohm metres (Ωm)

resistor an electronic component designed to introduce a known value of resistance into a circuit

root-mean-square (RMS) value RMS value is 0.707 of an ac voltage or current

rotor resistance starter a three-phase wound rotor induction motor controller which connects external resistances into the rotor circuit on starting and cuts them out when running

semi-enclosed fuse (BS 3036) rewirable fuse with tinned copper fuse wire

series dc machine a dc motor or generator with field windings and armature connected in series

shunt a parallel connected resistor used to extend the range of a moving coil ammeter to read values of current higher than the instrument's movement is designed for

shunt dc machine a dc motor or generator with field windings and armature connected in parallel

slip speed the difference between rotor speed (Nr) and synchronous speed (Ns)

soft-start a solid-state motor controller for starting a three-phase induction motor

solar power sunlight converted into electrical power

solenoid a coil consisting of a number of turns of wire wound in the same direction, which is capable of carrying a current

star a three-phase winding connection where the current through each phase winding I_P is the same as the line current I_L

star-delta starter a three-phase ac induction motor controller that connects the stator windings in star on starting and delta when the motor is running

statistics the study of collecting, sorting, analysing and presenting data

stroboscopic effect flicker from a discharge lamp can cause stroboscopic effect; a rotating machine may appear to be stationary or running slow when in fact it is still running at high speeds due to this flicker

super-grid system the HV transmission system operating at 275 or 400 kV

synchronous motor a single or three-phase motor that runs at synchronous speed

thermistor a temperature sensitive resistor whose resistance changes rapidly at a certain temperature switching point

thyristor a three terminal semiconductor device which is triggered into conduction by a pulse at its gate terminal

transformer an electrical machine that usually steps up or down an ac voltage and current from one circuit to one or more other circuits

transistor a three terminal semiconductor device used as an amplifier or a solid- state switching device

transposition of formula rearranging a formula to solve a problem (by finding an unknown variable)

triac a semiconductor device, similar to a thyristor, but which will conduct in either direction when a trigger pulse is applied to its gate terminal

universal motor a single-phase ac motor with field windings and armature connected in series, similar to a dc series wound motor

variable frequency drive a motor controller which allows a single or three-phase ac induction motor to run at speeds above synchronous speed

velocity *(v)* a measure of the speed of an object in a given direction, measured in metres per second *(m/s)*

vector a variable (such as force) having both direction and magnitude

voltage drop in an electrical circuit which normally occurs when current flows through the circuit conductors (the greater the resistance of the circuit conductors, the higher the voltage drop)

volume *(V)* the amount of space of a three-dimensional object or the space that a gas or liquid occupies, measured in cubic metre *(m³)*

wave energy energy generated by the force of ocean waves

weight the force experienced by a body due to the earth's gravitation, measured in newtons *(N)*

wind energy energy generated by harnessing the wind, usually by windmills

work *(W)* is done whenever an object is moved by a force, measured in joules *(J)* or newton-metres *(Nm)*: 1 J = 1 Nm

wound rotor a rotor with windings as fitted to a three-phase ac induction motor

Zener diode a semiconductor device used for voltage regulation; when reverse biased it suddenly increases in conductivity at a specific breakdown voltage

Index: Electrical Installation Book Eight